Proceedings of the Second International Colloquium on Numerical Analysis

Proceedings of the Second International Colloquium on Numerical Analysis

Plovdiv, Bulgaria, 13 - 17 August 1993

Editors: D. Bainov and V. Covachev

///VSP///
Utrecht, The Netherlands, 1994

VSP BV
P.O. Box 346
3700 AH Zeist
The Netherlands

First published in 1994

ISBN 90-6764-168-5

Printed in The Netherlands by Koninklijke Wöhrmann, Zutphen

CONTENTS

Preface

The Second International Colloquium on Numerical Analysis was organized by UNESCO and the Mathematical Society of Japan, with the cooperation of the Association Suisse d'Informatique, the Canadian Mathematical Society, Hamburg University, Kyushu University, the London Mathematical Society, Plovdiv University "Paissii Hilendarski", Technical University - Berlin, Technical University - Plovdiv, Union of the Bulgarian Mathematicians -Plovdiv Section, the University of Texas -Pan American, and sponsored by the firm "Eureka". It was held on August 13-17, 1993, in Plovdiv, Bulgaria. The subsequent colloquia will take place each year from 13-17 August in Plovdiv, Bulgaria.

The address of the organizing committee is:

Stoyan Zlatev, Mathematical Faculty of the Plovdiv University,
Tsar Assen Str.24, Plovdiv 4000, Bulgaria.

The Editors

2nd Int. Coll. on Numerical Analysis, pp. 3-11
D. Bainov and V. Covachev (Eds)

SPLINES ON $\mathbb{S}^2$ FOR SMALL LATITUDES

Angel Arteaga Iriarte*, Luis Hernández Encinas** and Jaime Muñoz Masqué***

* CSIC-IMAFF, Serrano 123, E-28006 Madrid, Spain. Email: ceeai00@cc.csic.es
** Dpto. Didáctica de las Matemáticas. Universidad de Salamanca. Paseo Canalejas 169. E-37007-Salamanca, Spain. Email: encinas@a200.usal.es
*** CSIC-IEC, Serrano 144, E-28006 Madrid, Spain. Email: vctqu01@cc.csic.es

Abstract

A notion of approximate spline curves for regions of small latitude on the sphere is introduced. And some of their properties are studied.

Keywords: Curvature, Euler-Lagrange equations, Lagrangian density, sphere, splines

1 Introduction

Methods for computing interpolating splines in the plane, even non-linear ones, and the corresponding algorithms implemented in a computer program are nowadays well known. For example, suffice it to mention the paper by Malcolm [1].

Moreover, recently, the squared curvature functional has been deeply studied on the spaces of constant curvature by several authors ([2], [3], [4], [5]), also including its global aspects and the qualitative behaviour of the solutions by means of Marsden-Weinstein reduction techniques for Hamiltonian system, Palais-Smale theory, etc.

In this setting, Euler-Lagrange equations are usually given by showing the differential equations that the curvature and the torsion must satisfy, assuming the curve is parametrized by its arc length. These equations can be then integrated by using Jacobi elliptic functions.

This procedure is not very suitable for computational purposes, because it does not provide directly a parametrization for the position coordinates of the curve in the manifold we were considering. Only for the Euclidean plane and the Hyperbolic plane, Bryant and Griffiths [4] gave the solutions in explicit form. The spherical case, however, seems to be much more difficult to be solved in this way. Nevertheless, the spherical case is the most important in Applied Sciences. Actually, the necessity of fitting a smooth path to some spherical data is a frequent problem that arises in a variety of situations in Biology, Geology, Physics, etc. ([6], [7], [8]).

Because of this, in the last ten years many papers have appeared proposing approximate methods of interpolating spline curves in the sphere (see [9], [10], [11], and the references therein)

These methods are specially successful when the spherical data are not nearly antipodal.

Here we present a method intended to spherical data which are distributed on a small neighbourhood of a great circle. If we think of this great circle as being the equator of the sphere, then we impose on our data to have a small latitude, but nothing is assumed on their longitudes.

2 The functional $\mathcal{L}(\gamma) = \int_\gamma \kappa^2 ds$ on $\mathbb{S}^2$

We shall always consider the unit sphere $\mathbb{S}^2$ of $\mathbb{R}^3$ endowed with the Riemannian metric g induced from the Euclidean metric on $\mathbb{R}^3$; i.e.,

$$g = \left(dx^2 + dy^2 + dz^2\right)_{|\mathbb{S}^2},$$

and we shall denote by ∇ the Levi-Civita connection of g.

Let

$$\left.\begin{array}{rcl} x & = & \cos\theta\cos\varphi \\ y & = & \cos\theta\sin\varphi \\ z & = & \sin\theta \end{array}\right\} \qquad -\frac{\pi}{2} \leq \theta \leq \frac{\pi}{2}, \qquad 0 \leq \varphi \leq 2\pi,$$

be the standard parametrization of $\mathbb{S}^2$, where θ stands for latitude and φ stands for longitude.

Hence,

$$\begin{array}{l} g = d\theta^2 + (\cos^2\theta)\, d\varphi^2, \\ \nabla_{\frac{\partial}{\partial\theta}} \dfrac{\partial}{\partial\theta} = 0, \quad \nabla_{\frac{\partial}{\partial\theta}} \dfrac{\partial}{\partial\varphi} = -(\tan\theta)\dfrac{\partial}{\partial\varphi}, \quad \nabla_{\frac{\partial}{\partial\varphi}} \dfrac{\partial}{\partial\varphi} = (\sin\theta\cos\theta)\dfrac{\partial}{\partial\theta}. \end{array} \tag{1}$$

Let us consider the functional:

$$\mathcal{L} : C^\infty\left([a,b], \mathbb{S}^2\right) \longrightarrow \mathbb{R},$$

given by

$$\mathcal{L}(\gamma) = \int_a^b \kappa^2 ds, \tag{2}$$

where κ is the curvature of γ and ds is the arc length of the curve γ: $[a,b] \longrightarrow \mathbb{S}^2$.

The functional $\mathcal{L}$ is defined by the second order Lagrangian density $\mathcal{L}dt$ on $J^2(\mathbb{R}, \mathbb{S}^2)$ determined by the condition:

$$(\mathcal{L}\, dt)_{(j_t^2\gamma)} = \kappa^2(t)\, ds,$$

where $\kappa(t)$ is the curvature of γ at the point $\gamma(t)$.

Bellow, we shall write:

$$\begin{array}{rclrcl} \theta(t) & = & \theta(\gamma(t)), & \varphi(t) & = & \varphi(\gamma(t)), \\ \dot\theta & = & \dfrac{d\theta}{dt}, & \dot\varphi & = & \dfrac{d\varphi}{dt}, \\ \ddot\theta & = & \dfrac{d^2\theta}{dt^2} = \dfrac{d\dot\theta}{dt}, & \ddot\varphi & = & \dfrac{d^2\varphi}{dt^2} = \dfrac{d\dot\varphi}{dt}, \text{ etc.} \end{array}$$

Let us first compute the local expression of $\mathcal{L}$. By using the formulas (1) (for g and ∇ on $\mathbb{S}^2$, we have:

$$\begin{aligned} T &= \text{tangent vector to } \gamma = \dot\theta\frac{\partial}{\partial\theta} + \dot\varphi\frac{\partial}{\partial\varphi}, \\ U &= \text{ unitary tangent vector to } \gamma = \frac{1}{\mu}T, \end{aligned} \tag{3}$$

where $\mu = \sqrt{g(T,T)} = \sqrt{\dot\theta^2 + (\cos^2\theta)\,\dot\varphi^2}$.
Hence,

$$\begin{aligned} \kappa^2 &= |\nabla_U U|^2 = g(\nabla_U U, \nabla_U U)\,, \\ \nabla_U U &= \frac{1}{\mu}\nabla_T\left(\frac{1}{\mu}T\right) = \frac{1}{\mu}\left[-\frac{\dot\mu}{\mu}T + \frac{1}{\mu}\nabla_T T\right], \\ \nabla_T T &= \ddot\theta\frac{\partial}{\partial\theta} + \ddot\varphi\frac{\partial}{\partial\varphi} + \dot\theta\left[\dot\theta\nabla_{\frac{\partial}{\partial\theta}}\frac{\partial}{\partial\theta} + \dot\varphi\nabla_{\frac{\partial}{\partial\varphi}}\frac{\partial}{\partial\theta}\right] + \dot\varphi\left[\dot\theta\nabla_{\frac{\partial}{\partial\theta}}\frac{\partial}{\partial\varphi} + \dot\varphi\nabla_{\frac{\partial}{\partial\varphi}}\frac{\partial}{\partial\varphi}\right] \\ &= \left(\ddot\theta + \dot\varphi^2\sin\theta\cos\theta\right)\frac{\partial}{\partial\theta} + \left(\ddot\varphi - 2\dot\theta\dot\varphi\tan\theta\right)\frac{\partial}{\partial\varphi}. \end{aligned}$$

Substituting T by its value given in (3):

$$\begin{aligned} \nabla_U U &= -\frac{\dot\mu}{\mu^3}\left(\dot\theta\frac{\partial}{\partial\theta} + \dot\varphi\frac{\partial}{\partial\varphi}\right) + \frac{1}{\mu^2}\left[\left(\ddot\theta + \dot\varphi^2\sin\theta\cos\theta\right)\frac{\partial}{\partial\theta} + \left(\ddot\varphi - 2\dot\theta\dot\varphi\tan\theta\right)\frac{\partial}{\partial\varphi}\right] \\ &= \frac{1}{\mu^2}\left(\ddot\theta + \dot\varphi^2\sin\theta\cos\theta - \frac{\dot\mu}{\mu}\dot\theta\right)\frac{\partial}{\partial\theta} + \frac{1}{\mu^2}\left(\ddot\varphi - 2\dot\theta\dot\varphi\tan\theta - \frac{\dot\mu}{\mu}\dot\varphi\right)\frac{\partial}{\partial\varphi}. \\ \kappa^2 &= \frac{1}{\mu^4}\left(\ddot\theta + \dot\varphi^2\sin\theta\cos\theta - \frac{\dot\mu}{\mu}\dot\theta\right)^2 + \frac{\cos^2\theta}{\mu^4}\left(\ddot\varphi - 2\dot\theta\dot\varphi\tan\theta - \frac{\dot\mu}{\mu}\dot\varphi\right)^2, \\ \kappa^2 ds &= \frac{1}{\mu^3}\left[\left(\ddot\theta + \dot\varphi^2\sin\theta\cos\theta - \frac{\dot\mu}{\mu}\dot\theta\right)^2 + \cos^2\theta\left(\ddot\varphi - 2\dot\theta\dot\varphi\tan\theta - \frac{\dot\mu}{\mu}\dot\varphi\right)^2\right] dt. \end{aligned}$$

Consequently,

$$\mathcal{L} = \frac{1}{\mu^3}\left[\left(\ddot\theta + \dot\varphi^2\sin\theta\cos\theta - \frac{\dot\mu}{\mu}\dot\theta\right)^2 + \cos^2\theta\left(\ddot\varphi - 2\dot\theta\dot\varphi\tan\theta - \frac{\dot\mu}{\mu}\dot\varphi\right)^2\right]. \tag{4}$$

3 The non-parametrized extremals of the functional $\mathcal{L}$

Let us assume that the curve $\gamma : \mathbb{R} \longrightarrow \mathbb{S}^2$ is an inmersion, say $\dot\theta \neq 0$. Then, by setting

$$\varphi' = \frac{d\varphi}{d\theta}, \quad \varphi'' = \frac{d^2\varphi}{d\theta^2}, \text{ etc.},$$

we have:

$$\dot\varphi = \dot\theta\varphi', \quad \ddot\varphi = \ddot\theta\varphi' + \dot\theta^2\varphi'', \quad \mu^2 = \dot\theta^2\left(1 + \varphi'^2\cos^2\theta\right), \text{ and}$$

$$\frac{\dot\mu}{\mu} = \frac{\ddot\theta}{\dot\theta} + \dot\theta\frac{\varphi'\varphi''\cos^2\theta - \varphi'^2\sin\theta\cos\theta}{1 + \varphi'^2\cos^2\theta}.$$

Substituting these expressions in the formula (4) for $\mathcal{L}$ obtained in the previous section, we have:

$$\mathcal{L}dt = \mathcal{L}\,'d\theta,$$

where

$$\begin{aligned}\mathcal{L}' &= \frac{[\varphi'^2 \sin\theta\cos\theta\,(2+\varphi'^2\cos^2\theta) - \varphi'\varphi''\cos^2\theta]^2}{(1+\varphi'^2\cos^2\theta)^{\frac{7}{2}}} \\ &+ \frac{\cos^2\theta\,[(\varphi''-2\varphi'\tan\theta)\,(1+\varphi'^2\cos^2\theta) - \varphi'^2\varphi''\cos^2\theta + \varphi'^3\sin\theta\cos\theta]^2}{(1+\varphi'^2\cos^2\theta)^{\frac{7}{2}}}\end{aligned}$$

The true spline curves in $\mathbb{S}^2$ should be defined as the extremal of the Lagrangian density $\mathcal{L}dt$; i.e., the solutions to the Euler-Lagrange equations

$$\left\{\begin{aligned}\frac{\partial\mathcal{L}}{\partial\theta} - \frac{d}{dt}\left(\frac{\partial\mathcal{L}}{\partial\dot\theta}\right) + \frac{d^2}{dt^2}\left(\frac{\partial\mathcal{L}}{\partial\ddot\theta}\right) &= 0\\ \frac{\partial\mathcal{L}}{\partial\varphi} - \frac{d}{dt}\left(\frac{\partial\mathcal{L}}{\partial\dot\varphi}\right) + \frac{d^2}{dt^2}\left(\frac{\partial\mathcal{L}}{\partial\ddot\varphi}\right) &= 0\end{aligned}\right.$$

Moreover, the formula $\mathcal{L}dt = \mathcal{L}'d\theta$ shows that spline curves are also extremals of the Lagrangian density $\mathcal{L}'d\theta$. More precisely: spline curves are the extremals of $\mathcal{L}'d\theta$ parametrized in a suitable way.

In what follows we shall only study non-parametrized spline curves; i.e., the extremals of $\mathcal{L}'d\theta$.

Lagrangian $\mathcal{L}'$ can be written as follows:

$$\begin{aligned}\mathcal{L}' &= \frac{\cos^2\theta}{(1+\varphi'^2\cos^2\theta)^{\frac{5}{2}}}\varphi''^2 - 2\varphi'\sin\theta\cos\theta\left[\frac{1}{(1+\varphi'^2\cos^2\theta)^{\frac{3}{2}}} + \frac{1}{(1+\varphi'^2\cos^2\theta)^{\frac{5}{2}}}\right]\varphi'' \\ &+ \varphi'^2\sin^2\theta\left[\frac{1}{(1+\varphi'^2\cos^2\theta)^{\frac{1}{2}}} + \frac{2}{(1+\varphi'^2\cos^2\theta)^{\frac{3}{2}}} + \frac{1}{(1+\varphi'^2\cos^2\theta)^{\frac{5}{2}}}\right].\end{aligned} \qquad (5)$$

This expression suggests to do the following transformation: $u = \varphi'\cos\theta$. Hence,

$$\frac{\partial\mathcal{L}'}{\partial\varphi''} = 2\frac{u'\cos\theta + u\sin\theta}{(1+u^2)^{\frac{5}{2}}} - 2u\sin\theta\left[\frac{1}{(1+u^2)^{\frac{3}{2}}} + \frac{1}{(1+u^2)^{\frac{5}{2}}}\right],$$

$$\begin{aligned}\frac{d}{d\theta}\left(\frac{\partial\mathcal{L}'}{\partial\varphi''}\right) &= \frac{-10uu'^2\cos\theta}{(1+u^2)^{\frac{7}{2}}} + \frac{2u''\cos\theta - 2u'\sin\theta + 6u^2u'\sin\theta}{(1+u^2)^{\frac{5}{2}}} \\ &- \frac{2u'\sin\theta + 2u\cos\theta}{(1+u^2)^{\frac{3}{2}}},\end{aligned} \qquad (6)$$

$$\begin{aligned}\frac{\partial\mathcal{L}'}{\partial\varphi'} &= \frac{-5uu'\cos\theta}{(1+u^2)^{\frac{7}{2}}} + \frac{2u'\sin\theta\,(3u^2-1)}{(1+u^2)^{\frac{5}{2}}} + \frac{(2-u^2)\,u\sin^2\theta - u'\sin 2\theta}{\cos\theta\,(1+u^2)^{\frac{3}{2}}} \\ &+ \frac{2u\sin^2\theta}{\cos\theta\,(1+u^2)^{\frac{1}{2}}}.\end{aligned} \qquad (7)$$

As $\dfrac{\partial\mathcal{L}'}{\partial\varphi} = 0$, the Euler-Lagrange equation for $\mathcal{L}'$ can be written in the form:

$$\frac{d}{d\theta}\left(\frac{\partial\mathcal{L}'}{\partial\varphi''}\right) - \frac{\partial\mathcal{L}'}{\partial\varphi'} = 2a, \qquad (8)$$

for some constant $a \in \mathbb{R}$, and the given formulas (6) and (7), yield:

$$u'' = \frac{5uu'^2}{2(1+u^2)} - \frac{u(u^2\sin^2\theta - 2)}{2\cos^2\theta}\left(1+u^2\right) + \frac{u\sin^2\theta}{\cos^2\theta}\left(1+u^2\right)^2 + \frac{a}{\cos\theta}\left(1+u^2\right)^{\frac{5}{2}}. \quad (9)$$

By subtracting $\frac{5uu'^2}{2(1+u^2)} + u(1+u^2) + a(1+u^2)^{\frac{5}{2}}$ to both sides of the above equation (9), we obtain:

$$u'' - \frac{5uu'^2}{2(1+u^2)} - u\left(1+u^2\right) - a\left(1+u^2\right)^{\frac{5}{2}}$$
$$= \frac{u(1+u^2)^2 + 3u(1+u^2)\sin^2\theta}{2\cos^2\theta} + a\left(1+u^2\right)^{\frac{5}{2}}\frac{2\sin^2\left(\frac{\theta}{2}\right)}{\cos\theta} \quad (10)$$

Moreover, for $|\theta| \le \frac{\pi}{4}$ we have $\cos\theta \ge \frac{1}{\sqrt{2}}$ and hence, $\left|\frac{\sin\theta}{\theta}\right| = \frac{\sin\theta}{\theta} < \frac{1}{\cos\theta} \le \sqrt{2}$. Therefore: $|\sin\theta| < \sqrt{2}\,|\theta|$. Consequently, for $|\theta| \le \frac{\pi}{4}$ we obtain:

$$\left|\frac{u'' - \frac{5uu'^2}{2(1+u^2)} - u(1+u^2) - a(1+u^2)^{\frac{5}{2}}}{(1+u^2)^{\frac{5}{2}}}\right|$$
$$\le \sin^2\theta\left(\frac{|u|}{\sqrt{1+u^2}} + \frac{3|u|}{(1+u^2)\sqrt{1+u^2}}\right) + 2\sqrt{2}\,|a|\sin^2\left(\frac{\theta}{2}\right) < \left(8+\sqrt{2}\,|a|\right)\theta^2.$$

In particular, for $|\theta| < \epsilon$ we have:

$$\left|\frac{u'' - \frac{5uu'^2}{2(1+u^2)} - u(1+u^2) - a(1+u^2)^{\frac{5}{2}}}{(1+u^2)^{\frac{5}{2}}}\right| < \left(8+\sqrt{2}\,|a|\right)\epsilon^2 \quad (11)$$

4 The approximating splines

The above relationship (11) is a differential inequality for small enough latitudes which is satisfied for every spherical spline. Accordingly, in dealing with small latitudes, in the first instance it is natural to consider the exact solutions to that differential inequality.

That is, letting $\epsilon = 0$, we find that equation (10) is equivalent to

$$u'' = \frac{5uu'^2}{2(1+u^2)} + u\left(1+u^2\right) + a\left(1+u^2\right)^{\frac{5}{2}}, \quad (12)$$

which is the differential equation of the approximating splines for small latitudes.

Note that we are led to the same result by letting $\theta \cong 0$ in the initial differential equation (9). As $u = \varphi'\cos\theta$, in the present case we can consider u as being a good approximation for φ'; i.e., $u \cong \varphi'$.

We set $p = u'$. Hence,

$$u'' = \frac{du'}{d\theta} = \frac{du'}{du}\frac{du}{d\theta} = \frac{dp}{du}p.$$

Therefore,

$$p\frac{dp}{du} = \frac{5up^2}{2(1+u^2)} + u\left(1+u^2\right) + a\left(1+u^2\right)^{\frac{5}{2}}, \tag{13}$$

which is a differential equation of first order.

We have:

$$\frac{d\left(p^2\right)}{du} = \frac{5u}{1+u^2}p^2 + 2u\left(1+u^2\right) + 2a\left(1+u^2\right)^{\frac{5}{2}}, \tag{14}$$

which is a linear differential equation on p^2.

Consequently, for some $b \in \mathbb{R}$:

$$p^2 e^{-\int \frac{5u}{1+u^2}du} - \int \left[2u\left(1+u^2\right) + 2a\left(1+u^2\right)^{\frac{5}{2}}\right] e^{-\int \frac{5u}{1+u^2}du} du = b.$$

Hence,

$$\frac{p^2}{(1+u^2)^{\frac{5}{2}}} - 2\int \left[\frac{u}{(1+u^2)^{\frac{3}{2}}} + a\right] du = b,$$

or else,

$$u'^2 = (b+2au)\left(1+u^2\right)^{\frac{5}{2}} - 2\left(1+u^2\right)^2,$$

that is,

$$u'^2 = \left(1+u^2\right)^2\left[(b+2au)\left(1+u^2\right)^{\frac{1}{2}} - 2\right]$$

and

$$u' = \left(1+u^2\right)\sqrt{(b+2au)\sqrt{1+u^2} - 2}.$$

Hence,

$$\frac{du}{(1+u^2)\sqrt{(b+2au)\sqrt{1+u^2}-2}} = d\theta$$

and

$$F(u) = \int \frac{du}{(1+u^2)\sqrt{(b+2au)\sqrt{1+u^2}-2}} = \theta - \theta_0 \tag{15}$$

As $u = \dfrac{d\varphi}{d\theta} = F^{-1}(\theta - \theta_0)$, we have:

$$\varphi - \varphi_0 = \int F^{-1}(\theta - \theta_0)\, d\theta \tag{16}$$

5 An example

Let us set

$$r_{a,b}(u) = \frac{1}{(1+u^2)\sqrt{(b+2au)\sqrt{1+u^2}-2}}$$

and let $a = 0, b = 3$.

We compute a list of values of u such that all the correspondent values of θ are less than $\frac{\pi}{9}$, for example:

$$u_i = \left\{-0.3 + i \times 75 \times 10^{-4}\right\}_{i=0,\dots,80} = \{-0.3, -0.2925, \dots, 0.2925, 0.3\}$$

If for each value of u, we compute the values of the following expression:

$$\theta_i = \left\{ \int_0^{u_i} r_{0,3}(u)\, du \right\}_{i=0,\dots,80},$$

we obtain a set of values of θ.

Moreover, we can compute for each value of θ the corresponding value of φ as follows:

$$\varphi_i = \left\{ \int_0^{\theta_i} u d\theta \right\}_{i=0,\dots,80} = \left\{ \int_0^{u_i} u r_{0,3}(u)\, du \right\}_{i=0,\dots,80}$$

Hence, we obtain for each value of u of the list a value for θ and another for φ:

u_i	θ_i	φ_i
−0.3000	−0.28547	0.04176
−0.2925	−0.27899	0.03984
...	...	...
−0.0075	−0.00749	0.00002
0	0	0
0.0075	0.00749	0.00002
...	...	...
0.2925	0.27899	0.03984
0.3000	0.28547	0.04176

To plot these points on $\mathbb{S}^2$ using *Autocad* software on a personal computer, we need transform these set of points into Cartesian coordinates. Hence, we transform the above set into another new one by computing:

$$\{x_i, y_i, z_i\}_{i=0,\dots,80} = \{\cos\theta_i \cos\varphi_i, \cos\theta_i \sin\varphi_i, \sin\theta_i\}_{i=0,\dots,80}$$

and we obtain the list of points in Cartesian coordinates:

x_i	y_i	z_i
0.95869	0.04006	−0.28161
0.96057	0.03828	−0.27538
...	...	...
0.99997	0.00002	−0.00749
1	0	0
0.99997	0.00002	0.00749
...	...	...
0.96057	0.03828	0.27538
0.95869	0.04006	0.28161

In this way we obtain a set of points on $\mathbb{S}^2$ and if we join each point with the following one by means of a segment, we can plot this set of points as a curve on $\mathbb{S}^2$.

Repeating this process for different couples of values for a and b, we can plot different sets of curves. Several examples of these curves can be seen in figures 1 and 2.

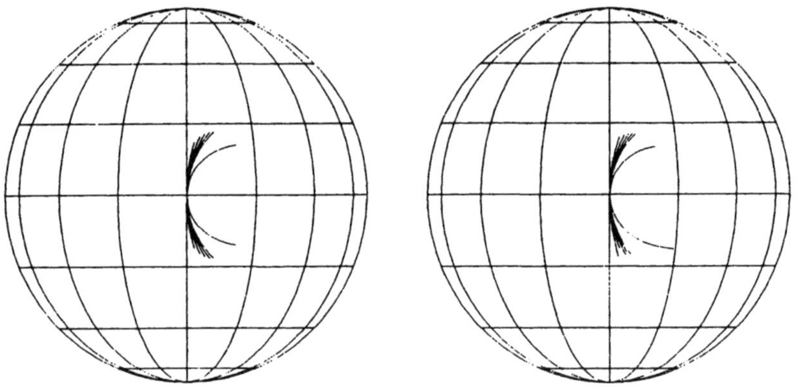

Figure 1: The curves for $a = 0$ and $a = 1$, with $b = \{3, 4, 5, 6, 7, 20\}$

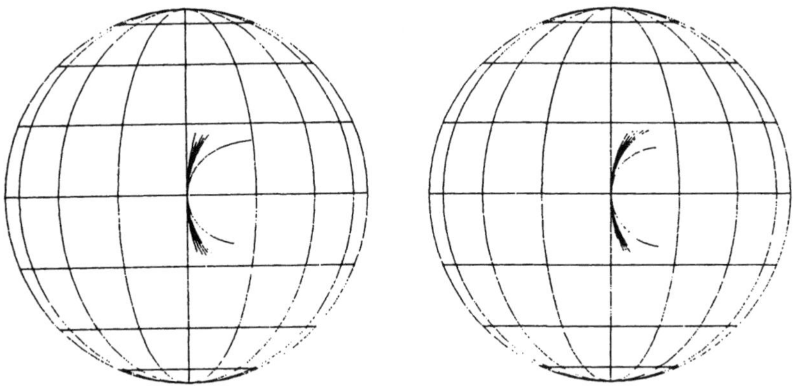

Figure 2: The curves for $a = -1$, with $b = \{3, 4, 5, 6, 7, 20\}$, and $a = 2$, with $b = \{4, 5, 6, 7, 8, 20\}$

References

[1] M.A. Malcolm, Siam J. Numer. Anal., 14, No. 2, 254-282 (1977).

[2] J. Langer and D.A. Singer, J. London Math. Soc. (2), 30, 512-520 (1984).

[3] J. Langer and D.A. Singer, J. Differential Geometry, 20, 1-22 (1984).

[4] R. Bryant and P. Griffiths, Amer. Journal of Math., 108, 525-570 (1986).

[5] J. Langer and D.A. Singer, Ann. Global Anal. Geom., 5, No. 2, 133-150 (1987).

[6] A.L. Gould, Biometrics, 25, 683-700 (1969).

[7] R.L. Parker and Ch.R. Denham, Geophys. J. R. Astr. Soc., 58, 685-687 (1979).

[8] R.M. Clark and R. Thompson, Geophys. J. R. Astr. Soc., 76, 337-368 (1984).

[9] N.I. Fisher and T. Lewis, J. R. Statist. Soc. B, 47, No. 3, 482-488 (1985).

[10] P.E. Jupp and J.T. Kent, Appl. Statist., 36, No. 1, 34-46 (1987).

[11] J. Hoschek and G. Seeman, Mathematical Modelling and Numerical Analysis, 26, No. 1, 1-22 (1992).

2nd Int. Coll. on Numerical Analysis, pp. 13-23
D. Bainov and V. Covachev (Eds)

The Algebraic Multilevel Iteration Methods - theory and applications

O. AXELSSON and M. NEYTCHEVA
Faculty of Mathematics and Informatics, Catholic University of Nijmegen, Toernooiveld, 6525 ED Nijmegen, The Netherlands

Abstract - A survey of the recently developed family of algebraic multilevel iteration methods is presented. The theory of the methods, together with the classes of problems considered, is exposed following the chronology of its derivation. This includes the computational complexity of the methods. A summary of some of the applications and illustrating numerical experiments are briefly described as well.

Keywords: Preconditioned conjugate gradient, selfadjoint and nonselfadjoint problems, optimal order preconditioners, computational complexity.

1. INTRODUCTION

When discretizing selfadjoint and positive definite elliptic problems applying a finite difference or finite element approach, we face the problem to solve linear systems of equations with symmetric positive definite matrices which are, in addition, large and sparse. The last two properties of the matrices, in general, imply that the iterative methods are the most effective solution methods for the above systems, and among them, the preconditioned conjugate gradient (PCG) method is mostly the best choice for problems with symmetric and positive definite matrices. Due to the well-known fact that the rate of convergence of the PCG method depends on the properties of the preconditioned matrix, much research work has been done to find "good" preconditioners to certain classes of matrices. For example, if a matrix, which is spectrally equivalent to the original one, is used as a preconditioner during the PCG iterations, then the rate of convergence becomes independent of the size of the problem. Such preconditioners are said to have optimal order with respect to the rate of convergence (property 'O1'). Another property tightly connected with a "good" preconditioner is, naturally, the low computational complexity of solving a system with it. The most desirable case would be to have an "optimal preconditioner with respect to the arithmetic (and communication, if applicable) work per degree of freedom per iteration step" (property 'O2'). Here we remark that it is not by mere chance that "good" is quoted. As it turns out a "good preconditioner" does not always coincide with the "optimal preconditioner" in some of the above meanings. For example, the spectral equivalence might not be of importance for the preconditioner if the solution of systems with it turns out to be very cheap on, say, a special type of computer architectures.

In this presentation we discuss, however, the algebraic multilevel iteration (AMLI) methods to construct preconditioners which possess both properties 'O1' and 'O2' and at the same time are suitable for classical sequential computers as well as for vector and (massively) parallel computer architectures.

The AMLI methods are based on a sequence of matrices $\{A^k\}$ corresponding to a given *nested* sequence of meshes or, more generally, matrix graphs.

For the nested discretizations, one way to construct optimal or nearly optimal preconditioners is to use the multigrid technique. A survey of some fundamental results for multigrid methods can be found in [1] and the references therein can be studied for more details. We point out that frecuently the optimal properties are achieved only if certain conditions hold, concerning the differential operator and the regularity of the region. Another difficulty with the multigrid methods is that they are not parameter-free and the proper choice of the parameters might be not an easy task to solve in more general situations. There are recent results, however, showing that only weak or no regularity assumptions are required (see [2], for example).

The AMLI method is based on purely algebraic means, which do not require the restrictions on the difference operator and the discretization as in the multigrid method. As its predecessors, the two level methods and the methods using the hierarchical basis functions can be considered (see, for example [1] and the references therein). Initially, nested meshes were considered (as in [3], [4], [5], for example), which were later generalized to nested matrix graphs. Some basic principles and results on multilevel preconditioners in a general framework are described and summarized in [6].

The ideas of the AMLI methods have also been applied for the class of nonselfadjoint and indefinite problems, where the preconditioned generalized conjugate gradient (GCG) method is a competitive alternative among the existing nonsymmetric linear solvers.

The rest of the paper is organized as follows. Section 2 contains the chronological description of the development of the AMLI methods. Subsection 2.1 is devoted to the results obtained for selfadjoint and positive definite problems. Different possibilities of constructing the AMLI preconditioner are briefly exposed together with the basic mathematical tools used while deriving the corresponding theory. In Subsection 2.2 some AMLI methods for nonselfadjoint and indefinite problems are described. The computational complexity of the AMLI methods is the topic of Section 3. Examples of applications and numerical results are given in Section 4. The conclusions are to be found in Section 5.

2. CHRONOLOGY OF THE DEVELOPMENT OF THE AMLI METHODS

2.1. *Selfadjoint second order elliptic problems*

Selfadjoint second order elliptic boundary value problems are considered in the major AMLI papers, such as [3], [4], [5], [7], [8], [9]. The constructed preconditioners can be applied to symmetric positive definite matrices, obtained after finite element (or finite difference) discretizations of the original problem.

2.1.1. *General description of the construction of the AMLI preconditioner.* As it is already stated, the AMLI methods are based on a sequence of matrices $\{A^{(k)}\}$. These matrices correspond to a given or defined nested sequence of discretizations (meshes) or matrix graphs so that the corresponding indices are taken from subsets N_k, for which

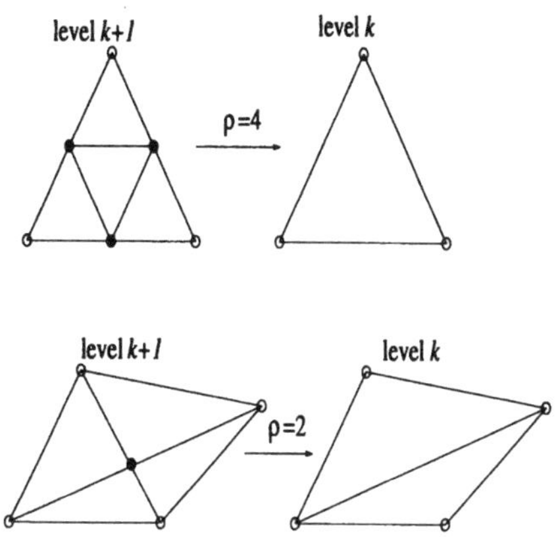

Figure 1. Examples of possible strategies for defining the AMLI levels.

$$N_0 \subset N_1 \subset \ldots \subset N_k \subset N_{k+1} \subset \ldots \subset N_\ell$$

holds. In the case of nested meshes, the simplest and most natural choice is to consider

$$A^{(k)} = \left[A_{ij}^{(k)}\right] = \left\{a(\phi_i^{(k)}, \phi_j^{(k)})\right\}_{(i,j)=1}^{n_k}, \tag{2.1}$$

where $a(\cdot,\cdot)$ is the corresponding elliptic bilinear form and $\{\phi_i^{(k)}\}$ is the set of finite element basis functions on N_k. This is studied in [3], [4], [9], for example.

More generally $A^{(k)}$ is defined approximating the Schur complement of $A^{(k+1)}$, computed for some partitioning and reordering of the entries of the latter, which does not necessarily correspond to certain discretization levels. This is considered, for example, in [8], where $A^{(k)}$ is defined as

$$A^{(k)} = A_{22}^{(k+1)} - A_{21}^{(k+1)} F_{11}^{(k+1)} A_{12}^{(k+1)},$$

and $F_{11}^{(k+1)}$ is some sparse, positive definite, nonnegative and symmetric approximation of $A_{11}^{(k+1)^{-1}}$. In such a case some extra conditions are imposed such as on the order n_k of the matrices $A^{(k)}$ which should decrease as a geometric ratio:

$$\frac{n_{k+1}}{n_k} = \rho_k \geq \rho > 1.$$

Figure 1 gives examples how ρ can be related to the different finite element meshes and the consecutive refinements.

After the sets N_k have been defined, each matrix $A^{(k+1)}$, $k > 0$ is permuted and partitioned in a two-by-two block matrix form as follows:

$$A^{(k+1)} = \begin{bmatrix} A_{11}^{(k+1)} & A_{12}^{(k+1)} \\ A_{21}^{(k+1)} & A_{22}^{(k+1)} \end{bmatrix} \begin{matrix} \} \, N_{k+1} \backslash N_k \\ \} \, N_k \end{matrix} . \tag{2.2}$$

Based on $\left\{A^{(k)}\right\}$ and the block partitioning (2.2), a sequence of implicitly defined preconditioners $\left\{M^{(k)}\right\}$ is then constructed, which become spectrally equivalent to $A^{(k)}$ under certain conditions.

We first consider the following definition for the preconditioning matrix, which is used in [3]:

$$M^{(0)} = A^{(0)}$$

$$\text{for } k = 0, 1, \ldots \ell{-}1$$

$$M^{(k+1)} = \begin{bmatrix} A_{11}^{(k+1)} & 0 \\ A_{21}^{(k+1)} & \widetilde{S}^{(k)} \end{bmatrix} \begin{bmatrix} I & A_{11}^{(k+1)^{-1}} A_{12}^{(k+1)} \\ 0 & I \end{bmatrix}, \tag{2.3}$$

where $\widetilde{S}^{(k)}$ can be either of the following:

$$\widetilde{S}^{(k)} = S^{(k+1)} \left[I - P_\nu (M^{(k)^{-1}} S^{(k+1)}) \right]^{-1}, \tag{2.4}$$

or

$$\widetilde{S}^{(k)} = A^{(k)} \left[I - P_\nu (M^{(k)^{-1}} A^{(k)}) \right]^{-1}. \tag{2.5}$$

The matrix $S^{(k+1)}$ in (2.4) is the exact Schur complement of $A^{(k+1)}$. $P_\nu(t)$ in the formulas (2.4) and (2.5) denotes a polynomial of degree ν which satisfies the conditions

$$0 \leq P_\nu(t) < 1,\ 0 < t \leq 1 \text{ and } P_\nu(0) = 1. \tag{2.6}$$

More details about the choice of the polynomial $P_\nu(t)$ follow in a forthcoming subsection.

In any iterative method with a preconditioner $M = M^{(\ell)}$, we need to solve systems with it. Note that the above definition of $M^{(\ell)}$ is implicit and it involves a solution of systems with $M^{(\ell-1)}$, which itself involves solutions with $M^{(\ell-2)}$ and so on. Due to the factorized form of $M^{(k+1)}$, the solution of a system $M^{(k+1)}\mathbf{x} = \mathbf{y}$ consists of two parts.

Forward sweep. Solve:

$$\begin{bmatrix} A_{11}^{(k+1)} & 0 \\ A_{21}^{(k+1)} & \widetilde{S}^{(k)} \end{bmatrix} \begin{bmatrix} \mathbf{w}_1 \\ \mathbf{w}_2 \end{bmatrix} = \begin{bmatrix} \mathbf{y}_1 \\ \mathbf{y}_2 \end{bmatrix}, \quad \text{i.e.} \quad \begin{array}{lrcl} (F1) & \mathbf{w}_1 & = & A_{11}^{(k+1)^{-1}} \mathbf{y}_1, \\ (F2) & \mathbf{w}_2 & = & \widetilde{S}^{(k)^{-1}} \left(\mathbf{y}_2 - A_{21}^{(k+1)} \mathbf{w}_1 \right). \end{array}$$

Backward sweep. Solve:

$$\begin{bmatrix} I & A_{11}^{(k+1)^{-1}} A_{12}^{(k+1)} \\ 0 & I \end{bmatrix} \begin{bmatrix} \mathbf{x}_1 \\ \mathbf{x}_2 \end{bmatrix} = \begin{bmatrix} \mathbf{w}_1 \\ \mathbf{w}_2 \end{bmatrix}, \quad \text{i.e.} \quad \begin{array}{lrcl} (B1) & \mathbf{x}_2 & = & \mathbf{w}_2, \\ (B2) & \mathbf{x}_1 & = & \mathbf{w}_1 - A_{11}^{(k+1)^{-1}} A_{12}^{(k+1)} \mathbf{x}_2. \end{array}$$

Denote $P_\nu(t) = 1 - a_1 t - \ldots - a_\nu t^\nu$. We observe that

$$\begin{aligned} \mathbf{v} = \widetilde{S}^{(k)^{-1}} \mathbf{z} &= \left[I - P_\nu (M^{(k)^{-1}} A^{(k)}) \right] A^{(k)^{-1}} \mathbf{z} \\ &= \left[a_1 I + a_2 M^{(k)^{-1}} A^{(k)} + \cdots + a_\nu (M^{(k)^{-1}} A^{(k)})^{\nu-1} \right] M^{(k)^{-1}} \mathbf{z} \end{aligned}.$$

Then the computation of $\mathbf{v} = \widetilde{S}^{(k)^{-1}} \mathbf{z}$ for some $\mathbf{z} \in \mathbf{R}^{n_k}$ can be implemented, for example for $\widetilde{S}^{(k)}$ defined by (2.5), in the following way:

Solve $M^{(k)}\mathbf{v} = a_\nu \mathbf{z}$.
For $r = 1$ step 1 until $\nu - 1$:
solve $M^{(k)}\mathbf{v} = A^{(k)}\mathbf{v} + a_{\nu-r}\mathbf{z}$.

Thus, the solution of a system with $M^{(k+1)}$ requires ν solutions with $M^{(k)}$ plus some matrix-vector multiplications and vector additions. The solution with $M^{(k)}$ itself involves ν solutions with $M^{(k-1)}$ ecetera.

It is obvious from (2.4) and (2.5) that the higher degree ν of the polynomials we choose, the better the approximation of the corresponding Schur complement $\widetilde{S}^{(k)}$ becomes and the more accurate solution is to be computed. On the other hand, the lower value of ν we use, the cheaper the method is. As shall be commented further, it is possible to compromise between the above two contradicting aims in the sense that there exist values of ν, for which properties 'O1' and 'O2' are simultaneously valid.

We analyze now the observation that at steps (F1) and (B2) we have to solve also systems with $A_{11}^{(k+1)}$. When $A^{(k)}$ is defined by (2.1), the condition number of the matrices $A_{11}^{(k+1)}$ is $O(1)$ and it is independent of k, then the solution can be done by some iterative method in an optimal order of computational complexity as accurate as we want. The cost, however, can be high and we have to do it twice. To overcome this, the next idea is undertaken, namely to replace $A_{11}^{(k+1)}$ with an approximation $B_{11}^{(k+1)}$ of it such that the solution with $B_{11}^{(k+1)}$ can be easier (cheaper) and still accurate enough. This possibility is studied in detail in [4], where instead of (2.3) the following preconditioner is defined:

$$M^{(k+1)} = \begin{bmatrix} B_{11}^{(k+1)} & 0 \\ \widetilde{A}_{21}^{(k+1)} & \widetilde{S}^{(k)} \end{bmatrix} \begin{bmatrix} I & B_{11}^{(k+1)^{-1}} \widetilde{A}_{12}^{(k+1)} \\ 0 & I \end{bmatrix} \tag{2.7}$$

with $\widetilde{S}^{(k)}$ - defined as in (2.5). The matrices $\widetilde{A}_{12}^{(k+1)}$ and $\widetilde{A}_{21}^{(k+1)}$ are easy to be computed as functions of the matrices $A_{11}^{(k+1)}$, $B_{11}^{(k+1)}$, $A_{12}^{(k+1)}$, $A_{21}^{(k+1)}$, J_{12} and J_{12}^T, respectively. The two matrices J_{12} and J_{12}^T define the connection between the stiffness matrices computed using the hierarchical two-level basis functions and those computed using the standard nodal basis functions.

Here, $B_{11}^{(k+1)}$ is an approximation of $A_{11}^{(k+1)}$ in the sense that there exists a positive constant δ_k so that

$$\mathbf{v}^T A_{11}^{(k+1)} \mathbf{v} \leq \mathbf{v}^T B_{11}^{(k+1)} \mathbf{v} \leq (1+\delta_k) \mathbf{v}^T A_{11}^{(k+1)} \mathbf{v} \text{ for any } \mathbf{v} \in \mathbf{R}^{n_{k+1}-n_k}. \tag{2.8}$$

We observe that for $\delta_k = 0$, $B_{11}^{(k+1)} \equiv A_{11}^{(k+1)}$ and (2.7) includes (2.3). In general $B_{11}^{(k+1)}$ can be some incomplete factorization of $A_{11}^{(k+1)}$. Naturally, the rate of convergence depends on the accuracy of the approximation – i.e., of the level dependent constant δ_k in (2.8).

Next step, dealt with in [5], is to avoid the dependency of the constant δ_k. It is shown that there is no need to estimate how accurate $B_{11}^{(k+1)}$ approximate $A_{11}^{(k+1)}$ on each level of refinement and still have the optimal rate of convergence property 'O2'.

As further step in the direction to simplify the task for solving systems with $M^{(\ell)}$, the preconditioner studied in [8] can be considered. Here, the recursive definition of $M^{(k+1)}$ is as follows:

$$\text{for } k = 0, 1, \ldots \ell{-}1$$
$$M^{(k+1)} = \begin{bmatrix} F_{11}^{(k+1)^{-1}} & 0 \\ A_{21}^{(k+1)} & I \end{bmatrix} \begin{bmatrix} I & F_{11}^{(k+1)} A_{12}^{(k+1)} \\ 0 & \widetilde{S}^{(k)} \end{bmatrix}, \tag{2.9}$$

where $\widetilde{S}^{(k)}$ is as in (2.5). This time the inverse of $A_{11}^{(k+1)}$ is approximated with some other matrix $F_{11}^{(k+1)}$. Thus, instead of solving systems with $A_{11}^{(k+1)}$ or with some approximation of it, we only have to do matrix multiplications with $F_{11}^{(k+1)}$. $F_{11}^{(k+1)}$ is constructed so that it meets some requirements such that it has a prescribed sparsity pattern, is nonnegative when $A_{11}^{(k+1)}$ is monotone and is a good enough approximation of $A_{11}^{(k+1)^{-1}}$. Its properties are described in detail in [10] and [8]. Some methods to construct such approximations can also be found in [8] and in the references therein.

2.1.2. *The role of the matrix polynomial.* It is seen from (2.4), (2.5) and (2.8) that the inverses of the Schur complements of the matrices $A^{(k)}$ - partitioned as in (2.2) - are approximated by certain matrix polynomials P_ν of degree ν, which satisfy the conditions (2.6). There exist, of course, a whole class of polynomials, which fulfill the conditions (2.6). For example,

$$P_\nu(t) = (1-t)^\nu$$

is such a one. The preconditioner obtained for the latter polynomial with $\nu = 1$ is one of the earliest, studied in [11].

In what follows, we add a subscript to the polynomial degree ν - ν_k to express the possibility that it may vary from one level to another.

There exist polynomials, however, for which an optimal rate of convergence is most easily achieved. These are the shifted and normalized Chebyshev type polynomials

$$P_{\nu_k}(t) = \frac{T_{\nu_k}\left(\frac{\beta_k+\alpha_k-2t}{\beta_k-\alpha_k}\right)+1}{T_{\nu_k}\left(\frac{\beta_k+\alpha_k}{\beta_k-\alpha_k}\right)+1}, \tag{2.10}$$

where $T_s(t)$ are the Chebyshev polynomials of degree s, $T_0 = 1$, $T_1 = t$, $T_{s+1} = 2tT_s - T_{s-1}$. Here, $0 < \alpha_k < \beta_k$ define an interval, which contains the spectrum of the level matrix product $M^{(k)^{-1}}A^{(k)}$. $P_{\nu_k}(t)$ defined in (2.10) have the smallest local maximum in $[\alpha, \beta]$, $0 < \alpha < \beta$.

It has been shown that there exist values of ν_k such that the relative condition number of $M^{(k)}$ with respect to $A^{(k)}$ is bounded independently of the level number k, thus the AMLI preconditioners possess the property 'O1'.

Initially, e.g. in [3] and [4], the polynomial degree ν is a parameter which is presumed to have one and the same value on each level. Clearly, $\nu = 1$ corresponds to the so-called 'V-cycle', which gives the cheapest possible preconditioner within the considered context. Later on, in [9], [10] and [8] it is shown that ν can differ from one level to another (in particular, to be greater than 1 after each group of k_0 levels) and still the multilevel preconditioner can possess the optimal property 'O1'. We postpone the discussion how the polynomial coefficients can be computed to a later section.

2.1.3. *Nonlinear (variable-step) preconditioners.* It is clear from the multilevel preconditioners, defined by (2.3), (2.7) and (2.8), that when the polynomial (2.10) is used, certain eigenvalue information at each level is required. The latter is needed for the computation of the coefficients for the polynomials $P_{\nu_k}(t)$ so that an optimal rate of convergence will be achieved. When symmetric positive definite matrices are considered, the necessary eigenvalue estimations are computed relatively easy. Despite the above, it still requires some extra computational efforts. Further, for nonsymmetric problems this approach becomes even more complicated.

One possible way to avoid eigenvalue estimates and also to be able to apply the multilevel preconditioners to nonsymmetric problems is presented in [7]. Here, more general interpretation of the action of the matrix polynomials at each level (or at each group of k_0 levels) is perceived, namely as an inner iteration method which solves a system of the type $\widetilde{S}^{(k)}\mathbf{v} = \mathbf{z}$. Then, the idea to replace the particular inner iteration process with another iterative procedure is developed. In [7], the generalized conjugate gradient method is used as an inner iteration method. In this way the proposed method becomes completely parameter-free. Symmetric and coercive bilinear forms with corresponding coefficient matrices with bounded and measurable entries are considered. The method is derived for finite element stiffness matrices $A^{(k)}$, computed form the bilinear form using the kth level hierarchical basis functions. Now a general description of the idea follows.

Let a system with a matrix A have to be solved with some iterative method, called "outer". To simplify the presentation, we consider the typical case when A is partitioned into two-by-two block form as in (2.2) and A_{11} is nonsingular. It can be seen that if $S = A_{22} - A_{21}A_{11}^{-1}A_{12}$, then

$$A^{-1} = \begin{bmatrix} A_{11} & A_{12} \\ A_{21} & A_{22} \end{bmatrix}^{-1} = \begin{bmatrix} I & -A_{11}^{-1}A_{12} \\ 0 & I \end{bmatrix} \begin{bmatrix} A_{11}^{-1} & 0 \\ -S^{-1}A_{21}A_{11}^{-1} & S^{-1} \end{bmatrix}. \qquad (2.11)$$

The action of A^{-1} on a vector involves two actions with A_{11}^{-1} and one with S^{-1}. We now replace in (2.11) all the actions of inverses of matrices by approximate actions, using some iterative method, called "inner". Then, the resulting matrix takes the form

$$M = \begin{bmatrix} I & -M_1[A_{11}^{-1}]A_{12} \\ 0 & I \end{bmatrix} \begin{bmatrix} M_2[A_{11}^{-1}] & 0 \\ -M_3[S^{-1}]A_{21}M_2[A_{11}^{-1}] & M_3[S^{-1}] \end{bmatrix}. \qquad (2.12)$$

and is used as a preconditioning matrix for the outer iterative method. Here, $M_s[Q]$, $s = 1,2,3$ denotes some approximate actions of a matrix Q. We have the freedom to use different approximate actions of one and the same matrix, to use less costly (and, perhaps, less accurate) actions during the first outer iterations than during the final ones. But from (2.12) it is clearly seen that even if A is symmetric, M may not be symmetric, in general. That is why GCG-type of solver should be used for the outer iterations.

In analogous way (cf. [7]) a multilevel variable-step preconditioner can be defined. For the latter, convergence properties are derived and it is shown that the GCG with the above preconditioner converges with optimal rate and optimal complexity. Estimates for 2-D as well as for 3-D problems are specified.

2.1.4. *Basic mathematical tools and theoretical results.* The main mathematical tool used in [3], [4], [5], [6] is the strengthened Cauchy-Bunyakowski-Schwarz (C.-B.-S.) inequality, which is of the following form for matrices partitioned as in (2.2):

$$| \mathbf{v}_1^T A_{12} \mathbf{v}_1 | = | \mathbf{v}_2^T A_{21} \mathbf{v}_2 | \leq \gamma \left(| \mathbf{v}_1^T A_{11} \mathbf{v}_1 | \right)^{1/2} \left(| \mathbf{v}_2^T A_{22} \mathbf{v}_2 | \right)^{1/2} \qquad (2.13)$$

for all $\mathbf{v}_1 \in \mathbf{R}^{n_{k+1}-n_k}$, $\mathbf{v}_2 \in \mathbf{R}^{n_k}$ and with $0 < \gamma < 1$. Here, in the finite element context, hierarchical basis functions are assumed.

The conditions for spectral equivalence between $M^{(\ell)}$ and $A^{(\ell)}$ to hold are in terms of lower bounds of the degree ν of the matrix polynomial involved in the construction of $M^{(\ell)}$, which are independent of ℓ. The major result states that ν, when it is fixed on each level, must be chosen so that

$$\nu > \left(1 - \gamma^2\right)^{-(1/2)},$$

where γ is the constant in the strengthened C.-B.-S. inequality (2.13) for the corresponding two-level hierarchical basis function spaces, thus, independent of the regularity of the solution. As one can find in [12] for instance, γ can be easily computed from the local finite element matrices. The value of the upper bound β_k for the coefficients in (2.10) is computed as a function of γ and α_k in these cases is 1.

In [9] the lower bound for ν is in terms of the growth of the energy norm $\eta(k_0)$ of the nodal interpolation operator, restricting a function from a finer element space to a coarser one. Optimal order of complexity results for 2-D and 3-D are derived.

The optimality properties 'O1' and 'O2' of the multilevel preconditioner (2.9), described in [8], are based on the assumption that $A^{(\ell)}$ is a Stieltjes matrix and on the properties of the approximation matrix to $A_{11}^{(k+1)^{-1}}$, $k = \ell - 1, \cdots, 0$. Here, the main result is of the form

$$\rho^{\mu+1} > \nu > F(\beta_k),$$

where the function F is the max of the products of β_k within the k_0 groups of levels and $\nu_k = 1$ on all levels except at the end of each cycle of k_0 levels, where $\nu_k = \nu$. In this case β_k is computed using a few Lanczos iterations, combined with a bisection method. The values of α_k are found using a simple recursive formula, involving α_{k-1}.

2.2. *Nonselfadjoint and indefinite problems*

An application of the variable-step preconditioner concept from [7] to the generalized conjugate gradient method (proposed in [13]) is derived in [14]. Here, the solution of systems of linear equations $A\mathbf{x} = \mathbf{b}$ with possibly nonsymmetric and/or indefinite matrix A is considered. A variable-step preconditioner, seen also as a, in general, nonlinear mapping $B[\cdot] : \mathbf{r} \rightarrow B[\mathbf{r}]$ is used to accelerate the global convergence of the GCG method. Under assumptions of coercivity and boundedness of the mapping $B[\cdot]$, a monotone convergence of the GCG method with variable-step preconditioner is proved. Conditions for the coercivity and boundedness of the mapping $B[\cdot]$ to hold are also derived.

More about the theory of multilevel preconditioning indefinite elliptic finite element matrices can be found in [14], [16] and [17].

3. COMPUTATIONAL COMPLEXITY ASPECTS

We restrict ourselves to comment on the computational complexity of AMLI only for symmetric positive definite problems. As already stated for the AMLI methods considered so far, they have

'O1' an optimal rate of convergence, i.e. – independent of the level number;

'O2' an optimal order of computational complexity, i.e. – proportional to the degrees of freedom (meshpoints) on the finest level.

Even though 'O1' and 'O2', the following question arises quite naturally. Could other methods, even with nonoptimal order, give in total less computational labor than that needed for AMLI? Major part of the research work done to answer the above question can be found in [18]. Here, estimates for the asymptotic computational work and also for the rate of convergence are derived. The figures show that the AMLI implementation requires only a bounded number of arithmetic operations per meshpoint, which is independent of the number of levels as well as of the coefficients of the differential operator. It can depend, however, on the initial triangulation in the finite element mesh used. For more details and further discussions, see the references in [18] and also [9], [8], [19], [20] and [21].

When implementing AMLI on parallel architectures, the communication time factor has also to be taken into account. Going "down" to more coarse meshes we have to deal with both the effects of (a) increasing distances to communicate and (b) decreasing order of matrix and vector objects to work on. Because of the above *it is advisable to choose the coarsest mesh fairly fine.* Since the method uses only matrix-vector multiplications, vector additions and scalar products (for the PCG) its implementation on shared memory machines can be quite efficient, specially after the above consideration has been taken into account.

For distributed memory machines with p processors, the scalar products require global communications which, in general, are performed in $O(\log p)$ steps. On hypercube architectures it is efficient to map the nodepoints onto the processing elements using a binary reflected Gray code ordering. Then, if the distance between two node points is a power of 2 (which is the usual case if the nodes belong to different meshes), the actual communication distance for them becomes 2. We remark, however, that even though the latter is a prerequisite to expect very good efficiency in such cases, the communication time depends on the distance as well as on the virtual ratio, i.e. the number of nodepoints mapped onto one physical processor. Some theoretical reasonings and examples of parallel implementations of AMLI on hypercube architectures (CM-2 and CM-200) are given in [21], [22] and [23].

4. APPLICATIONS

Different AMLI preconditioners have been applied to various problems, showing the optimal behavior of the proposed method with respect to properties 'O1' and 'O2'. We briefly mention some of the examples.

Problem 1. $-\nabla \mathbf{k} \nabla u = f$, $\mathbf{x}$ in $\Omega \subset \mathbf{R}$, $u = 0$ on $\partial\Omega$.

Problem 2. $-\Delta u = f$ in Ω, Dirichlet and Neumann boundary conditions.

Problem 3. $\begin{array}{l} -div(k(x,y)\nabla u) + \mathbf{p}\nabla u + qu = f \text{ in } \Omega, \\ u = 0 \text{ on } \partial\Omega,\ 0 < c_1 < k(x,y) < c_2 \end{array}$.

Problem 4. $-div(k(x,y)\nabla u) - \lambda qu = f$ in Ω, $u = g$ on $\partial\Omega$.

Problem 5. $-(au_x)_x - (bu_y)_y = f$ in Ω, Dirichlet, Neumann boundary conditions.

Problem 1 is considered in [3], discretized by nested sequence of triangular meshes and piecewise linear finite element basis functions. The coefficients can be discontinuous but they are assumed to be constant on each element on the coarsest triangulation. The domain Ω can be rectangular or L-shaped region with reentrant corner. The same differential problem is tested in [7] and [9]. Problem 2 is the test problem in [4], [5] on an L-shaped domain and in [8] on a square region. In [18] domain decomposition preconditioners with inexact subdomain solvers are tested on Problem 3. Here, the discretization matrix is nonsymmetric. The properties of the preconditioners for nonsymmetric and indefinite problems in [16] are tested on Problem 3 and Problem 4, respectively. The convection-diffusion problem is considered on a polygonal domain, discretized using mixed finite elements. Numerical examples, based on Problem 5 (strong anisotropy), are analyzed in [24]. Tests with large values of the ratio a/b or b/a are coupled with discontinuity of the coefficients. An optimal order multilevel preconditioner of the type (2.3) — (2.5) is constructed so that the solution algorithm does not depend on the ratio of the anisotropy. In [25] AMLI is used within the context of solving the generalized symmetric eigenvalue problem.

5. CONCLUSIONS

The AMLI methods can be analyzed both as

- methods for a sequence of finite element or finite difference nested meshes, or
- general approximate factorization methods, which can be stabilized to an arbitrary good condition number for symmetric positive definite problems. This can be done within a purely algebraic general framework.

Acknowledgement

This research has been partly supported by The Netherlands Organization for Scientific Research N.W.O. under Grant no. 611-302-021.

References

[1] O. Axelsson and P.S. Vassilevski, A survey of multilevel preconditioned iterative methods, *BIT*, 29, 769-793, 1989.

[2] J.H. Bramble, J.E. Pasciak, J. Wang and J. Xu, Convergence estimates for multigrid algorithms without regularity assumptions, *Math. Comp.*, 57, 23-45 (1991).

[3] O. Axelsson and P.S. Vassilevski, Algebraic multilevel preconditioning methods I, *Numer. Math.*, 56, 157-177 (1989).

[4] O. Axelsson and P.S. Vassilevski, Algebraic multilevel preconditioning methods II, *SIAM J. Numer. Anal.*, 27, 1569-1590 (1990).

[5] O. Axelsson and P.S. Vassilevski, Algebraic multilevel preconditioning methods III, Report 9045, October 1990, Department of Mathematics, Catholic University, Toernooiveld, 6525 ED Nijmegen, The Netherlands.

[6] O. Axelsson, An algebraic framework for hierarchical basis functions multilevel methods or the search for 'optimal' preconditioners. In D. Kincaid and L. Hayes, editors, *Iterative Methods for Large Linear Systems*, pages 17-40, Academic Press, Inc., 1990.

[7] O. Axelsson and P.S. Vassilevski, Variable-step multilevel preconditioning methods, I: Selfadjoint and positive definite elliptic problems, *Num. Lin. Alg. Appl.*, to appear.

[8] O. Axelsson and M. Neytcheva, Algebraic multilevel iteration method for Stieltjes matrices, *Num. Lin. Alg. Appl.*, to appear.

[9] P.S. Vassilevski, Hybrid V-cycle algebraic multilevel preconditioners, *Math. Comp.*, 58, 489-512 (1992).

[10] O. Axelsson, The method of diagonal compensation of reduced matrix entries and multilevel iterations, *J. Comp. Appl. Math.*, 38, 31-43 (1991).

[11] P.S. Vassilevski, Nearly optimal iterative methods for solving finite element equations based on multilevel splitting of the matrix, Report *no.* 1989-09, Institute for Scientific Computation, University of Wyoming, Laramie, Wyoming, 1989.

[12] V. Eijkhout abd P.S. Vassilevski, The role of the strengthened Cauchy-Bunyakowski-Schwarz inequality in multilevel methods, *SIAM Review*, 33, 405-419 (1991).

[13] O. Axelsson, A generalized conjugate gradient, least square method, *Numer. Math.*, 51, 209-227 (1987).

[14] O. Axelsson and P.S. Vassilevski, A black-box generalized conjugate gradient solver with inner iterations and variable-step preconditioning, *SIAM J. Matrix Anal. Appl*, 12, 625-644 (1991).

[15] O. Axelsson and P.S. Vassilevski, Construction of variable-step preconditioners for inner-outer iteration methods. In R. Beauwens and P. de Groen, editors, *Iterative Methods in Linear Algebra*, Elsevier Science Publishers B.V. (North Holland), 1992.

[16] P.S. Vassilevski, Preconditioning nonsymmetric and indefinite finite element elliptic matrices, *J. Num. Lin. Alg. Appl.*, 1, 59-76 (1992).

[17] P.S. Vassilevski, Indefinite elliptic problem preconditioning, *Communications in Applied Num. Meth.*, 8, 257-264 (1992).

[18] O. Axelsson and P.S. Vassilevski, Asymptotic work estimates for AMLI methods, *Applied Numerical Mathematics*, 7, 437-451 (1991).

[19] O. Axelsson and V. Eijkhout, Analysis of a recursive 5-point/9-point factorization method. In O. Axelsson and L.Yu Kolotilina, editors, *Preconditioned Conjugate Gradient Methods*, number 1457 in Lecture Notes in Math., pages 154-173, Springer Verlag, 1990.

[20] O. Axelsson and V. Eijkhout, The nested recursive two-level factorization method for nine-point difference matrices, *SIAM J. Scientific and Statistical Computations*, 12, 1373-1400 (1991).

[21] V. Eijkhout, Implementation of 5-point/9-point multi-level methods on hypercube architecture, *Proceedings of the 1990 International Conference on Supercomputing, Amsterdam*, ACM press, 291-295 (1990).

[22] I.D. Mishev, V. Austel, T.F. Chan and P.S. Vassilevski, Experiments with algebraic multilevel preconditioners on Connection Machine, CAM Report 93-25, UCLA.

[23] M.G. Neytcheva, Implementing the AMLI algorithm on the Connection Machine, manuscript in progress.

[24] S.D. Margenov and P.S. Vassilevski, Algebraic Multilevel Preconditioning of Second Order Elliptic Problems with Strong Anisotropy, Report No. 07, 1991, University of Wyoming, Laramie, Wyoming, U.S.A.

[25] O. Axelsson and M. Neytcheva, Finding eigenvalues in an interval using parallelizable algorithms, Report 9302, January 1993, Department of Mathematics and Informatics, Catholic University, Toernooiveld, 6525 ED Nijmegen, The Netherlands.

[26] K. Stüben, Algebraic multigrid (AMG): experiences and comparisons. In S.F.McCormick, U.Trottenberg, editors, Appl. Math. Comp., *Proc. of the International Multigrid Conference*, Copper Mountain, CO, April 6-8 (1983).

2nd Int. Coll. on Numerical Analysis, pp. 25-34
D. Bainov and V. Covachev (Eds)

THE ALGEBRAIC ANALYSIS OF INCOMPLETE FACTORIZATION PRECONDITIONINGS

ROBERT BEAUWENS
Service de Métrologie Nucléaire, Université Libre de Bruxelles (C.P. 165)
50, av. F.D. Roosevelt, B-1050 Brussels, Belgium

Abstract

Our purpose here is to summarize the progress made by the algebraic analysis of incomplete factorization preconditionings used to solve (by preconditioned polynomially accelerated iterative methods) large sparse symmetric positive definite linear systems such as arising from the finite difference or finite element discretization of elliptic partial differential equations.

We first briefly review the past and present evolution of these methods since their discovery in the early sixties and their rediscovery in the early seventies to the sophisticated versions that are presently under investigation, stressing old hopes, basic difficulties and feasible issues that lead to present day methods.

Definitions and notation used throughout the paper are next explained together with the basic principles of both the geometrical and algebraic analyses.

The algebraic approach is then covered in more detail under its present version stressing the new issues that followed from this approach : the *perturbation* point of view, the development of *dynamic* factorization algorithms, the analysis of *block* modified incomplete factorizations. We shall also stress the role played by the *depth* of the graph of the upper triangular factor in the algebraic approach.

The more recent evolution of both the geometrical and algebraic approaches will then be broadly considered, stressing their common concern of specific ordering assumptions, stressing also that, once again, the geometrical approach is ahead (inasmuch as one can still make a distinction between them).

Key words : Iterative methods, preconditioning, incomplete factorizations, M-matrices, matrix graphs.

AMS(MOS) : 65F10, 65N20, 65F35, 15A23, 15A48.

1. Introduction

Our purpose in this talk is to review the basic results of the algebraic analysis of the so-called *modified incomplete* or MILU factorizations when used as preconditioners of polynomially accelerated iterative methods (Chebyshev, conjugate gradients, etc.). Although the ultimate purpose of an algebraic approach is the development of black box robust and performant iterative solvers, this cannot yet be considered as reached without having some more or less specific field of applications in mind. The field of applications behind the developments to be summarized here is the iterative solution of symmetric positive definite linear systems arising from finite difference or finite element discretizations of second order elliptic equations.

The introduction of both ILU and MILU preconditionings goes back to the early sixties [1, 2, 3] and it was soon argued by Price and Varga [4] that the rapidly increasing complexity of higher order ILU factorizations limited their applicability to low order methods with the corollary that ILU preconditionings were unable to improve the Jacobi preconditioning by an order of magnitude - as initially hoped - both preconditioning leading, for standard model problems, to similar asymptotic behaviours.

The Price-Varga argument remained unpublished but came to be known (or similar observations were done) in the early seventies, initiating research in other directions [5, 6, 7, 8, 9], a.o. on the SSOR and the MILU preconditionings. It should be mentioned that, meanwhile, the ILU methods had been simultaneously and independently rediscovered by Woznicki [10, 11] and by Meijerink and vand der Vorst [12, 13]. The last paper was very successful and the methods became widely used in spite of the Price-Varga argument.

The first significant progress in the conditioning analysis of MILU factorizations were obtained by the geometrical approach developed in the works of Axelsson [8] and Gustafsson [9]. These were basic contributions, to be considered as the first break through the $0(1/h^2)$ barrier set up by the Price-Varga argument.

The algebraic approach followed in the eighties [14, 15, 16, 17, 18] and was responsible for a series of developments, a.o. the introduction of the *perturbation* point of view (i.e. the distinction between *perturbed* and *unperturbed* MILU factorizations), the first analyses of *block* MILU factorizations [19, 20], the introduction of *dynamic* factorizations [15, 21], the extension of many results to the iterative solution of singular systems [22, 23].

This approach has seen many new developments in the recent years, too numerous to be completely reviewed here, but which can more or less all be considered as centered around the ordering problem, i.e. the definition of the most appropriate ordering(s) to reach both good conditioning and good parallel implementation.

It is however again by geometrical approaches that new basic progresses have been reached about this specific concern, arising a.o. from the use of so-called *hierarchical bases* [24] and of *repeated red-black* orderings [25]. We expect that these new results will soon be integrated by the algebraic approach, mainly because one of the basic lemmas of [25] is actually of algebraic type.

2. Background material

Let $A = (a_{ij})$ be a symmetric positive definite $n \times n$ matrix, $U = (u_{ij})$ an upper triangular matrix with positive diagonal entries, $P = diag(U)$ ($diag(A)$ denotes here the diagonal matrix with same diagonal entries as A and $offdiag(A)$ denotes $A - diag(A)$) and

$$B = U^T P^{-1} U \quad . \tag{1}$$

The matrix B may be considered as an approximate factorization of A provided that it is, in some sense, close to A and we shall now define a few relevant types of approximations. Note that the *existence* of each of these approximations is not a priori granted.

Ideally, one would like to have $B = A$, leading to

$$u_{ij} = a_{ij} - \sum_{k<i} \frac{u_{ki}u_{kj}}{u_{kk}} \quad , \quad i \leq i \leq j \leq n \ . \tag{2}$$

For large sparse matrices A, this is often prohibitively expensive, whence the need for inexact (sparse) approximations.

Incomplete factorizations may be defined by

$$u_{ij} = a_{ij} - \beta_{ij} \sum_{k<i} \frac{u_{ki} u_{kj}}{u_{kk}} \quad , \quad i \leq i \leq j \leq n \tag{3}$$

where $\beta_{ij} \in \{0,1\}$ denote a given set of *masks*. These approximations avoid excessive fill-in by simply ignoring (*masking*) it.

Modified incomplete factorizations may be defined by

$$u_{ij} = a_{ij} - \beta_{ij} \sum_{k<i} \frac{u_{ki} u_{kj}}{u_{kk}} \quad , \quad i \leq i < j \leq n \tag{4}$$

where $\beta_{ij} \in \{0,1\}$ denote a given set of (offdiagonal) masks, while the diagonal entries of U are chosen such as to satisfy the (generalized) row sum criterion

$$Bx = Ax + \Lambda\, Dx \tag{5}$$

where x is some given positive vector, $D = diag(A)$ and Λ is a nonnegative diagonal matrix of *small* parameters called here *perturbations.*

The purpose of the conditioning analysis is to provide rules for the choice of U such as to keep the spectral condition number

$$\kappa(B^{-1}A) = \max_{z \neq 0} \frac{(z, Az)}{(z, Bz)}$$

of the *preconditioned* matrix $B^{-1}A$ as small as possible.

The first tool of all analyses is a *reduction* technique based on the following (obvious) property : if A_o is a symmetric positive definite matrix related to A by

$$\forall z \neq 0 \, : \, \alpha \leq \frac{(z, Az)}{(z, A_o z)} \leq \beta \, ,$$

with $\alpha > 0$, then

$$\kappa(B^{-1}A) \leq \frac{\beta}{\alpha} \kappa(B^{-1}A_o) \, .$$

This technique *reduces* the problem of estimating $\kappa(B^{-1}A)$ to that of estimating $\kappa(B^{-1}A_o)$. It also introduces new approximate factorizations of A since B may now be defined as an incomplete or modified incomplete factorization of A_o. On the other hand, it raises the problem of estimating the parameters α and β and is (therefore) partly responsible of the limitations of the presently available analyses.

In the following we shall generally assume that A is a Stieltjes matrix (actually meaning hereby that it can be *reduced* to the Stieltjes case). This may be considered as the only limitation of the algebraic approach. Geometrical approaches on the other hand assume (more pragmatically) that A is the matrix of some precise discrete model problem (meaning hereby again that it can be *reduced* to one of these model matrices).

The geometrical approaches have therefore at their disposal a natural parameter (the mesh space h of the discrete model problem) in function of which they can express their results. To get comparable results, the algebraic approach rests on the use of graph topological properties, a few of which will now be briefly summarized. We use here directed rather than undirected graphs because this will enable us to state the basic results of the algebraic approach for general approximate factorizations, of the form $B = S^T P^{-1} S$ with $P = diag(S)$ without assuming that S is triangular. Lack of space will however prevent us from trying to take advantage of this more general setting.

A directed graph $G = (N, E)$ is an ordered pair of sets, the node set N and the edge set E, the elements (i,j) of which are ordered pairs of nodes $i,j \in N$. The graph $G(A) = (N, E(A))$ of an $n \times n$ matrix $A = (a_{ij})$ may be defined by

$$(i,j) \in E(A) \Leftrightarrow a_{ij} \neq 0 \tag{6}$$

using the set of the first n integers as node set.

A node j (of some graph $G = (N, E)$) is a *precursor* of i if $(j,i) \in E$; it is a *successor* of i if $(i,j) \in E$; the set of precursors (successors) of i is denoted by $P(i)$ $(S(i))$; the set of common precursors of a pair of nodes i,j is denoted by

$$Pc(i,j) = P(i) \cap P(j)$$

and we further let, for $M \subset N$,

$$Pc(M) = \bigcup_{\substack{i,j \in M \\ i \neq j}} Pc(i,j) ;$$

we also write $Pc(G)$ for $Pc(N)$ and, when G is the graph $G(A)$ of some $n \times n$ matrix A, we further write $Pc(A)$ for $Pc(G(A))$.

A path of length ℓ in a graph is a sequence of nodes $i_o, i_1, \ldots, i_\ell$ with $(i_k, i_{k+1}) \in E$; it is called simple if all nodes are distinct with the possible exception of the first and last ones, in which case it is a (simple) cycle. A *directed acyclic graph* or DAG is a directed graph without cycles. A square matrix can be permuted to (lower or upper) triangular form if and only if its graph is a DAG.

A node is called initial (final) if it has no precursors (successors); such nodes exist in any DAG; the *depth* of a DAG is the maximal length of its paths; if M is a subset of the node set of a DAG, we denote by $\ell(M)$ the depth of the subgraph induced by M.

The *ascent* $As(i)$ of a node i in a DAG is defined by

$$As(i) = \{j \mid \text{there exists a path from } j \text{ to } i\} .$$

3. The conditioning analysis

Our purpose here is to state (Theorem 1 below), prove and discuss one of the basic results of the algebraic approach to the conditioning analysis of a positive definite matrix A by an approximate factorization B. We do not attempt to put this result under its most general form because this would obscure the meaning of the result. For this reason, we shall a.o. assume that A is a Stieltjes matrix. On the other hand we first state an auxiliary lemma, the proof of which we skip because it is exactly the same as that of Lemma 1 in [26].

Lemma 1. Let S be an M-matrix with $P = diag(S)$, let x be a positive vector and define

$$t = \max_{i \in Pc(S)} \frac{((P-S)x)_i}{(Px)_i} . \tag{7}$$

Let further the matrices B_1 and B_2 be defined by

$$\begin{aligned} \mathit{offdiag}(B_1) &= \mathit{offdiag}(S^T + S) \ , \\ \mathit{offdiag}(B_2) &= -\mathit{offdiag}((S-P)^T P^{-1}(S-P)) \ , \\ B_1 x &= B_2 x = 0 \ . \end{aligned}$$

Then, the matrix $\tau B_1 - B_2$ is positive semidefinite for any $\tau \geq t$.

Theorem 1. Let A be a Stieltjes matrix, S an M-matrix, $P = diag(S)$, $B = S^T P^{-1} S$ and assume that

$$- \mathit{offdiag}((S-P)^T P^{-1}(S-P)) \leq \mathit{offdiag}(S^T + S - A) \leq 0 \ . \tag{8}$$

Let x be a positive vector, define t by (7) and assume further that $t < 1$ and that

$$(1-\tau)Ax \leq Bx \leq Ax \tag{9}$$

for some $t \leq \tau < 1$. Then, for any $z \neq 0$,

$$1 \leq \frac{(z, Az)}{(z, Bz)} \leq \frac{1}{1-\tau} \tag{10}$$

whence $\kappa(B^{-1}A) \leq 1/(1-\tau)$.

Proof. Letting the matrices B_1 and B_2 be defined as in Lemma 1, it is readily checked that

$$\mathit{offdiag}(B - (1-\tau)A - (\tau B_1 - B_2)) = \mathit{offdiag}((1-\tau)(S^T + S - A)) \leq 0$$

while

$$(B - (1-\tau)A - (\tau B_1 - B_2))x = Bx - (1-\tau)Ax \geq 0$$

implying that $B-(1-\tau)A-(\tau B_1 - B_2)$ is positive semidefinite. Since $\tau B_1 - B_2$ is positive semidefinite by Lemma 1, the right inequality (10) follows.

The left inequality follows in a similar way, by checking that $\mathit{offdiag}(A - B) \leq 0$ with $(A-B)x \geq 0$. □

We now discuss the scope of this result. The assumption (8) is always satisfied by *incomplete factorizations* of Stieltjes matrices and the best way to satisfy (9) is to compute – if possible – the diagonal entries of S such that

$$Bx = Ax$$

which is the rule (5) associated with unperturbed modified incomplete factorizations.

It is not always possible to execute this algorithm but, when S is triangular (or permutable to triangular form), the necessary and sufficient conditions under which it will succeed are known as well as the appropriate safeguards to insert against failures. These results will be briefly discussed in the next section. For more general approximate factorizations, where S cannot be permuted to a triangular matrix, the existence problem is open. We shall therefore assume in the following that the graph $G(S)$ of S is a DAG, i.e. that the unknowns can be reordered such that S is upper triangular.

The crucial assumption is then $t < 1$ and more precisely how big we can take $1 - \tau$, i.e. how large is $1 - t$. Note that $1 - t$ is a measure of the *degree* of (generalized) diagonal dominance of S (with respect to x and $Pc(S)$). The following result gives sufficient conditions to have such an estimate. It is a simplified version of Theorem 4.2 [18] to which we refer for a proof.

Theorem 2. Let A be a Stieltjes matrix, S an M-matrix whose graph $G(S)$ is a DAG of depth $\ell+2$, $P = diag(S)$, $B = S^T P^{-1} S$ and assume that

$$-\mathit{offdiag}((S-P)^T P^{-1}(S-P)) \leq \mathit{offdiag}(S^T + S - A) \leq 0 \ .$$

Let x be a positive vector such that $Bx = Ax \geq 0$ and assume that

$$(S^T - S)x \ \leq \ Bx \ . \tag{11}$$

Then, for any $z \neq 0$,

$$1 \ \leq \ \frac{(z, Az)}{(z, Bz)} \ \leq \ \ell + 2 \tag{12}$$

i.e. $\kappa(B^{-1}A) \leq \ell + 2$.

The notation $\ell+2$ for the depth of $G(S)$ in this theorem is intended to imply that it is at least 2, i.e. the minimal one compatible with $Pc(G) \neq \emptyset$. (In the trivial case where $Pc(G) = \emptyset$, the assumptions of the theorem actually imply that $B = A$).

To put this result into historical perspective, it should be mentioned that, by the time its original version appeared [14], it was the first one applicable with $Bx = Ax$. Former results by Axelsson's [8] and Gustafsson's [9] geometrical approaches assumed that

$$Bx \ \geq \ Ax + \xi h^2 \tag{13}$$

where h was the mesh size of the discrete model problem and $\xi > 0$, leading to

$$\kappa(B^{-1}A) \ \leq \ \frac{c^t}{\sqrt{\xi} h} \ . \tag{14}$$

Because this bound was diverging for small values of ξ, the additional *modification*, although quite small, had not been presented as a *perturbation*; Theorem 2 made it feasible to apply standard perturbation analysis [16] to the methods using (13) and the *perturbation* point of view readily became standard.

Comparing the bounds (12) and (14) it is seen that $\ell+2$ plays the role of the parameter $1/h$ of the geometrical approach. In the case of a rectangular grid with lexicographic ordering (as used in these geometrical approaches) one has indeed that $\ell + 2 \div 1/h$. In other words, one may consider the depth of the graph $G(S)$ as an algebraic substitute for the parameter $1/h$ of the geometrical approach; it has the interest of being always defined, even when there is no underlying discrete geometric problem.

Another issue of Theorem 1 (and other similar results) was the development of *dynamic* factorization algorithms. The idea was to enforce the condition $t < 1$ (since this is one of the key assumptions, directly governing the upper spectral bound) rather than $Bx = Ax$; the technique was to increase the diagonal entries of S, where necessary; the result was that the upper bound in (10) was a priori granted but $Bx = Ax$ was lost, exchanged for (5) with nonzero perturbation matrix Λ and the lower bound of (10) was – therefore – also lost. Detailed studies of specific examples [27, 16, 17] showed however that it was a viable solution and a heuristic formula was found [18] to estimate the new lower bound in all cases. A preliminary version of this dynamic algorithm may be found in [15] : Eq. (7.18); its refined definition was given in [21] : strategy 4.

Subsequently, other spectral bounds have been found resulting from the works of Axelsson, Eijkhout and Notay and they were used by Notay to set up a variety of new dynamic factorizations. We refer to [28] for a review. It seems presently that one of the most robust challenger to the dynamic method described above might be the so-called DRIC method of Notay [29].

Before concluding this section, we must notice that sparse block incomplete factorizations (see [30, 31, 19]) can also be implemented as equivalent point incomplete factorizations. Of course this is never done in practice because it would seriously increase the memory requirements and the computational complexity of the algorithm but it shows how to apply the preceding results to sparse block incomplete factorizations. This remark lead to the first analysis of block MILU methods in [20]. Subsequently, this subject was much more deeply analysed by Magolu [32, 33, 34, 35] who extended in particular the definition of dynamic factorization algorithms to the sparse block schemes.

4. The ordering problem

For a variety of motivations, including the searches for improved spectral conditioning and good parallel implementations, recent interest in approximate factorizations has centered around the ordering problem. Also seemingly unrelated work may be viewed under this perspective.

The experimental work by Duff and Meurant [36] is a good introduction to this topic. It shows a variety of orderings that have been suggested (far from exhaustive in spite of the large number of different orderings considered) and their influence on the convergence behaviour of the associated PCG methods with associated ILU or MILU preconditioning.

For MILU factorizations, the first relevant result to be mentioned in this field is the recent solution of the existence problem by Notay [23]; in the nonsingular case considered here, Notay's criterion may be stated as follows.

Theorem 3. Let A be a Stieltjes matrix, let $D = diag(A)$, let x be a positive vector such that $Ax \geq 0$ and let F be a nonnegative strictly upper triangular matrix such that

$$offdiag(F^T + F + A) \ \geq \ 0 \ .$$

Then, there exists a nonnegative nonsingular diagonal matrix P such that

$$(P - F)^T P^{-1} (P - F)x \ = \ Ax + \Lambda Dx$$

if and only if, for any node i in the graph $G(F)$ of F,

$$S(i) \ = \ \emptyset \ \Rightarrow \ \exists j \in As(i) \text{ with } (Ax)_j \ > \ 0 \ .$$

The application of this criterion to the Duff and Meurant experiments using unperturbed MILU preconditioning (cf. Table 5.6 of [36]) shows why the use of the orderings called there *mind, rle, altd*, and *4col* lead to divergent behaviours; the application of the algebraic analysis of the previous section also shows that small positive perturbations of the zero pivots would have been sufficient to make these methods convergent and further that the use of perturbed MILU preconditioning would have made it feasible to estimate $\kappa(B^{-1}A)$ for all orderings of this table.

Other works [37, 38, 39, 40, 26] added ordering assumptions to improve the analysis, essentially showing that, for a variety of orderings, the perturbations were unnecessary or could at least be substantially reduced. Further, all these orderings also display nice properties for the parallel implementation of MILU preconditioning.

The best conditioning results in this area came however from geometrical analyses, by Yserentant [24] and by Brand [25]. The work of Yserentant on *hierarchical basis multigrid* preconditioning does not seem at first sight to fit into our context; what it does however is an approximate factorization using a nonstandard ordering; the resulting preconditoned system has spectral condition number bounded by $O(log_2(1/h))$ (which by the way turns out to be the depth of the triangular factors and actually appears as such in Yserentant's analysis).

The work by Brand [25] is the analysis of unperturbed MILU preconditioning of the discrete Poisson problem on a square grid using *repeated red-black* ordering and it leads to a spectral condition number bounded by $O(1/\sqrt{h})$. Here again, $1/\sqrt{h}$ is the depth of the graph of the triangular factors.

Because the number of PCG iterations required for solving the problem to prescribed accuracy is bounded by $O(\sqrt{\kappa(B^{-1}A)})$, all these orderings often lead to comparable iteration numbers in actual practice. Further research should therefore concentrate on enlarging the scope of these analyses as well as investigating their practical implementations on vector and parallel computers.

Acknowledgement

This work presents research results of the Belgian Incentive Program "Information Technology" – Computer Science of the future, initiated by the Belgian State – Prime Minister's Service – Science Policy Office (Contract No. IT/IF/14). The scientific responsibility is assumed by its author.

References

[1] N. Buleev, Mat. Sb., 51, pp. 227–238 (1960). English translation in : Report BNL-TR-551, Brookhaven National Laboratory, Upton, New-York, 1973.

[2] R. Varga, in Boundary Problems in Differential Equations, R. Langer, ed., University of Wisconsin Press, Englewood Cliffs, pp. 121–142 (1960).

[3] T. Oliphant, Quart. Appl. Math., 20, pp. 257–267 (1962).

[4] H. Price and R. Varga, Incomplete primitive factorizations, unpublished manuscript (1964).

[5] H. Stone, SIAM J. Numer. Anal., 5, pp. 530–558 (1968).

[6] T. Dupont, R. Kendall, and H. Rachford, SIAM J. Numer. Anal., 5, pp. 559–573 (1968).

[7] D. Young, J. of Approx. Theory, 5, pp. 137–148 (1972).

[8] O. Axelsson, BIT, 13, pp. 443–467 (1972).

[9] I. Gustafsson, BIT, 18, pp. 142–156 (1978).

[10] Z. Woznicki, PhD thesis, Institute of Nuclear Research, Swierk, Poland, (1973).

[11] Z. Woznicki, Nukleonika, 23, pp. 941–968 (1978).

[12] J. Meijerink and H. van der Vorst, Technical Report, ACCU-Reeks, 11, Academisch Computer Centrum Utrecht, (1974).

[13] J. Meijerink and H. van der Vorst, Math. Comp., 31, pp. 148–162 (1977).

[14] R. Beauwens, Lin. Alg. Appl., 62, pp. 87–104 (1984).

[15] O. Axelsson and V. Barker, Finite Element Solution of Boundary Value Problems. Theory and Computation, Academic Press, New York (1984).

[16] R. Beauwens, Lin. Alg. Appl., 68, pp. 221–242 (1985).

[17] ——, Lin. Alg. Appl., 85, pp. 101–119 (1987).

[18] R. Beauwens and R. Wilmet, J. Comput. Appl. Math., 26, pp. 257–269 (1989).

[19] R. Beauwens and M. Ben Bouzid, SIAM J. Numer. Anal., 24, pp. 1066–1076 (1987).

[20] ——, SIAM J. Numer. Anal., 25 (1988), pp. 941–956.

[21] R. Beauwens, in Preconditioned Conjugate Gradient Methods, O. Axelsson & L. Kolotilina, eds., Lectures Notes in Mathematics No. 1457, Springer-Verlag (1990) pp. 1–16.

[22] Y. Notay, BIT, 29, pp. 682–702 (1989).

[23] ——, in Preconditioned Conjugate Gradient Methods, O. Axelsson and L. Kolotilina, eds., Lectures Notes in Mathematics No. 1457, Springer-Verlag (1990) pp. 105–125.

[24] H. Yserentant, Numer. Math., 49, pp. 379–412 (1986).

[25] C. W. Brand, Numer. Math., 61, pp. 433–454 (1992).

[26] R. Beauwens, Num. Lin. Alg. with Appl., 1 (1993).

[27] I. Gustafsson, Research Report 77.13R, Dept. of Computer Sciences, Chalmers Univ. of Technology and Univ. of Göteborg, Göteborg, Sweden (1977).

[28] R. Beauwens, 92 Shangai International Numerical Algebra and its Applications Meeting, Shangai, October 26-30, 1992.

[29] Y. Notay, A dynamic version of the RIC method, submitted for publication (1993).

[30] R. Beauwens, Series in Appl. Math., Report Nr. 74-1, Northwestern University Evanston, Illinois, (1974).

[31] P. Concus, G. Golub, and G. Meurant, SIAM J. Sci. Statist. Comput., 6, pp. 220–252 (1985).

[32] M. Magolu, Appl. Numer. Math., 8, pp. 25–42 (1991).

[33] ——, Numer. Math., 61, pp. 91–110 (1992).

[34] ——, in Iterative Methods in Linear Algebra, R. Beauwens and P. de Groen, eds., North-Holland (1992) pp. 519–529.

[35] ——, PhD thesis, Service de Métrologie Nucléaire, Université Libre de Bruxelles, Brussels, Belgium (1992).

[36] I. Duff and G. Meurant, BIT, 29, pp. 635–657 (1989).

[37] R. Beauwens, BIT, 29, pp. 658–681 (1989).

[38] Y. Notay, PhD thesis, Service de Métrologie Nucléaire, Université Libre de Bruxelles, Brussels, Belgium (1991).

[39] A. Kutcherov and M. Makarov, J. Numer. Lin. Alg. Appl., 1, pp. 1–26 (1992).

[40] R. Beauwens, Y. Notay, and B. Tombuyses, Num. Lin. Alg. with Appl., 1 (1993).

2nd Int. Coll. on Numerical Analysis, pp. 35-44
D. Bainov and V. Covachev (Eds.)

Numerical Computation of Incompressible Flows

Charles-Henri BRUNEAU
CeReMaB - Université Bordeaux I
33405 Talence (France)

ABSTRACT - After a short review of the various forms of Navier-Stokes equations for incompressible flows, this paper propose a whole numerical method to solve the classical formulation in velocity-pressure. The discretization is based on finite differences with reduction of numerical diffusion by adding an anti-diffusion term in space and by treating the convection terms explicitely in time. Results for the 2D driven cavity problem and the flow behind a cylinder in a channel are obtained for high Reynolds numbers. In the last case, open boundary conditions on artificial limits are discussed and a new condition in traction is tested successfully. Although we claim to compute solutions as accurate as possible, it is not yet sure that these solutions have a quantitative meaning.

1 - Introduction

These last two decades, solving incompressible Navier-Stokes equations has been probably one of the most exciting topics for numerical analysts. Moreover, with the growth of computing power, it is now possible to simulate directly the transition to the turbulence, at least in 2D. So, a wide range of problems is under investigation. Here, we focus on two classical problems in 2D, the driven cavity and the flow behind a cylinder in a channel which contain the main difficulties for open and close geometries as we shall see.

Trying to make an exhaustive review of all the work that has been done on the subject is out of reach even if some authors have done it quite well (see for instance [1]). Here, we describe only some aspects starting with the continuity equation

(1) $$div\, U = 0 \quad in \quad Q_T = \Omega \times (0, T)$$

where $U = (u_1, u_2, u_3)^T$ is the velocity, Ω the computational domain and $(0, T)$ the time interval the solution is observed.

This equation characterizing incompressible fluids is in most cases used in this form either as a constraint or as one of the equations to solve. For instance, it can be used to compute the pressure as it is done in our strongly coupled approach but also separately by adding a pressure term

$$\frac{\partial p}{\partial t} + div\, U = 0\,.$$

On the other hand, the momentum equation is written in various forms according to the choice of unknowns and the space dimension. Starting from the equation in velocity-pressure

(2) $$\frac{\partial U}{\partial t} + (U.\nabla)U - \frac{1}{Re}\,\Delta U + \nabla p = f \quad in \quad Q_T$$

where Re is the Reynolds number and f the external forces, or

(2') $$\frac{\partial U}{\partial t} + (U.\nabla)U - div\,\sigma(U,p) = f \quad in \quad Q_T$$

where $\sigma(U,p) = \frac{2}{Re}\,\varepsilon(U) - pI$ is the stress tensor with $\varepsilon(U)_{ij} = \frac{1}{2}\left(\frac{\partial u_i}{\partial x_j} + \frac{\partial u_j}{\partial x_i}\right)$, we can apply the divergence operator to get an elliptic equation for the pressure

$$\Delta p = div\, f - div((U.\nabla)U)\,.$$

For two dimensional spaces, one can use the famous stream function-vorticity formulation by setting

$$u_1 = \frac{\partial \psi}{\partial x_2}\;,\; u_2 = -\frac{\partial \psi}{\partial x_1}\;,\; \omega = \frac{\partial u_2}{\partial x_1} - \frac{\partial u_1}{\partial x_2} = -\,\Delta\psi$$

and by applying the curl operator on (2)

$$\frac{\partial \omega}{\partial t} + \frac{\partial}{\partial x}\left(\frac{\partial \psi\omega}{\partial y}\right) - \frac{\partial}{\partial y}\left(\frac{\partial \psi\omega}{\partial x}\right) - \frac{1}{Re}\,\Delta\omega = curl\, f\,.$$

Unfortunately, this form can not be extended in 3D and is replaced by the velocity-pressure form

$$\frac{\partial \omega}{\partial t} + (U.\nabla)\omega - (\omega.\nabla)U - \frac{1}{Re}\Delta\omega = curl\, f$$

where $\omega = curl\, U$ or in Poisson form $-\Delta U = curl\,\omega$.

For all these models of Navier-Stokes equations, discretization can be achieved by finite differences, finite elements, finite volumes or spectral methods in Fourier, Lagrange or Chebishev expansions. In addition, according to the way the continuity equation is treated, we can use a Lagrangien or penalty method, a decoupled or a strongly coupled approach. This explains the large amount of numerical methods proposed to the reader to solve incompressible Navier-Stokes equations. However, the main point is not to solve these equations for some easy or complex geometries and to get values of numerical solutions, but to get values close to the reality. Indeed, each of the numerous methods described in the literature gives numerical solutions but it is always difficult to know if these solutions are realistic. Let us take an example, many authors have worked on the 2D driven cavity problem ; for $Re = 100$ everyone gets the same solution, for $Re = 1000$ there are some small discrepancies between the numerical solutions and for

Re = 5000 or 10000 you can find all kind of results depending on the numerical viscosity and the robustness of the method. The solutions of incompressible Navier-Stokes equations are closely related to one parameter, the Reynolds number. Small changes on the value of Re can imply strong changes on the solution, the steady solution can lost its stability to the benefit of a periodic solution, a periodic solution can have a doubling of period and so on. So, we have to be very carefull if we want to compute accurate solutions. If the method is too weak, we can not get the transition to turbulence but if the method is too strong, it often contains too much numerical viscosity and the computed solutions do not correspond to the right value of the Reynolds number but to a smaller one. Thus, the art is to find the right balance between stability and viscosity to get the right solution. The last few years, several authors have proposed good scenarios for the transition to turbulence. According to the knowledge of dynamical systems, it seems that their solutions are qualitatively correct but we do not know if they have a quantitative meaning because of the numerical viscosity.

In this paper, we justify the choice of the discretization in space and time by the wish to compute solutions as close as possible to the exact solutions. We propose also open boundary conditions on artificial limits that yield a well-posed problem and give good numerical results. Finally, we present some numerical tests for both the driven cavity problem and the flow behind a cylinder in a channel with solid walls. For high Reynolds numbers, the behaviour of the numerical solutions is correct and we expect they are quantitatively good.

2 - Governing equations and boundary conditions

We consider the system of the equations (1) and (2), for the unknowns U and p, associated to the initial condition

$$U(x,0) = U_0(x) \quad in \quad \Omega \tag{3}$$

and boundary conditions.

For the driven cavity problem, the boundary conditions are of Dirichlet type as shown on figure 1

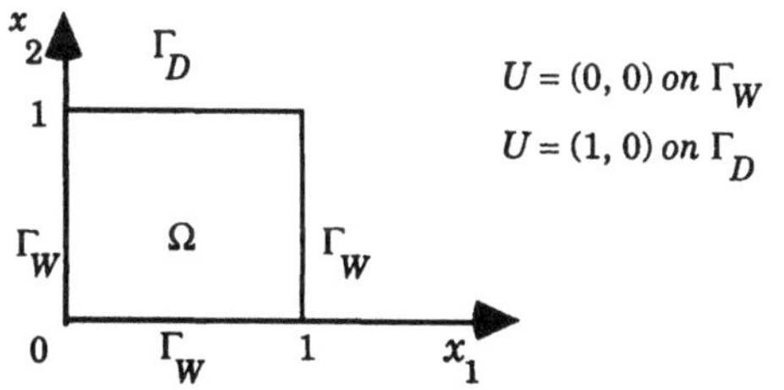

Figure 1. Domain and boundary conditions for the driven cavity problem.

For the flow behind a cylinder in a channel, the boundary conditions are more complex except on the walls where we always have a no-slip boundary condition (see figure 2)

(4) $$U=(0,0) \ \ on \ \ \Gamma_{W_0} \cup \Gamma_{W_1}$$

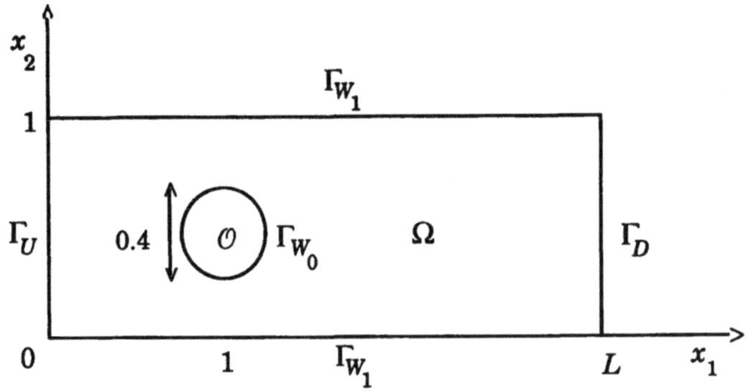

Figure 2. Domain and notations for the channel.

In addition, if the entrance section is not too close to the obstacle we can impose a Poiseuille flow, let us say with a flowrate of one :

(5) $$U = U_P = (6x_2(1-x_2), 0) \ \ on \ \ \Gamma_U$$

But on the exit section, as we can not put the boundary too far away from the obstacle, we need an open condition good enough to avoid reflections and to allow the vortices to cross the artificial boundary. The reader can find once again many such conditions in the literature. Among them, let us mention these based on the traction $F=\sigma(U,p)n$ or the pseudo-traction $\tilde{F}=\tilde{\sigma}(U,p)n$ where $\tilde{\sigma}(U,p)=\frac{1}{Re}\nabla U - pI$ and n represents always the unit normal vector pointing outside of the domain. For instance :

$$F = F_0 \ given$$
$$\tilde{F} = 0$$
$$F_n = 0 \ and \ U_n = u_0 \ given$$
$$or \ \ F + \alpha U = 0 \ \text{ for a chosen parameter } \alpha .$$

There are also conditions based on a parabolization of the Navier-Stokes equations as

$$\frac{\partial p}{\partial \tau} = -\frac{\partial U_\tau}{\partial t} - U_\tau \frac{\partial U_\tau}{\partial \tau} + \mu \frac{\partial^2 U_\tau}{\partial \tau^2}$$

$$\frac{\partial U_\tau}{\partial n} = 0$$

where (n, τ) forms an orthonormal basis, or conditions on the pressure as p=0 for Poiseuille flow, $p+\frac{1}{2}U^2=p_0$ for potential flow and $p=\frac{1}{Re}\frac{\partial U_n}{\partial n}-F_n$

for pressure Poisson model.

Here, we propose one of the natural boundary conditions, we have established for (2'), that yield a well-posed problem (see [2] for more details)

$$\sigma(U,p)n+\frac{1}{2}U_n^-(U-U_P)=\sigma(U_P\,,0)n \tag{6}$$

where we adopt the convention $a=a^+-a^-$ and we associate a zero constant pressure to Poiseuille flow downstream. When the flow is outgoing, this condition reduces to the traction is given equal to the traction of Poiseuille flow ; otherwise we take into account the convection terms to convey properly the vortices through the artificial limit.

In practise, this condition is coupled to the continuity equation to get three equations for the three unknowns. Let us write for the channel

$$\begin{aligned}&\frac{2}{Re}\frac{\partial u_1}{\partial x_1}-p+\frac{1}{2}\,u_1^-(u_1-6x_2\,(1-x_2))=0\\ &\frac{\partial u_1}{\partial x_2}+\frac{\partial u_2}{\partial x_1}+\frac{1}{2}\,u_1^-u_2=\frac{6}{Re}\,(1-2x_2) \qquad on\ \Gamma_D\\ &\frac{\partial u_1}{\partial x_1}+\frac{\partial u_2}{\partial x_2}=0\,.\end{aligned} \tag{6'}$$

Finally, for the channel, equations (1) - (2) are associated to the initial condition (3) and boundary conditions (4) - (5) - (6) or (6'). But it remains one difficulty that is the discretisation of the obstacle $\mathcal{O}$. As we want to use finite differences, we need a cartesian grid which is not convenient to approximate a cylinder. So, instead of solving Navier-Stokes equation (2) in the fluid domain Ω , we consider the obstacle as a porous media and solve the following Navier-Stokes-Darcy equation in the whole domain $(0,L)\times(0,1)$ as in [3] .

$$\frac{\partial U}{\partial t}+(U.\nabla)U-\frac{1}{Re}\,\Delta U+\frac{U}{ReDaK^3}+\frac{1}{K}\,\nabla p=0 \tag{2''}$$

where K is the permeability of the media with $K=K_F=1$ in Ω for the fluid and $K=K_S=10^{-8}$ in $\mathcal{O}$ for the solid, and Darcy number is given by $Da=\frac{1}{ReK_S}$ so that a slightly perturbed Navier-Stokes equation is solved in the fluid and a slightly perturbed Darcy equation is solved in the solid, both coupled to the continuity equation.

3 - Numerical method

We start this section by the discretization of equations (1) - (2) or (1) - (2") for the channel ; the additional term $\frac{U}{ReDaK^3}$ is a term of mass that gives a positive contribution on the main diagonal for $\frac{1}{ReDaK^3}$ is always positive.

For the sake of simplicity, we consider the 1D viscous Burgers equation to explain the finite differences approximation

$$\frac{\partial u}{\partial t} + u\,\frac{\partial u}{\partial x} - \nu\,\frac{\partial^2 u}{\partial x^2} = 0\,. \tag{7}$$

As in [4] , we use centered finite differences for the diffusion term and uncentered finite differences for the convection term based on Murman scheme and corrected by an antidiffusion term $\eta\ |u|\ \frac{\partial^2 u}{\partial x^2}$ with $\eta = min(\frac{\nu}{|u|}, \frac{\Delta x}{2})$ to insure the stability. Indeed, if the diffusion term is greater than the numerical viscosity term of Murman scheme, we suppress the artificial viscosity. Otherwise, we keep the necessary amount to get the stability. In the numerical tests we take $\eta = \frac{\Delta x}{3}$ that gives the best results as we have shown in [5] . For the time derivative, we need an accurate scheme if we want to follow correctly the solution. We recommend a third order Gear scheme with an explicit treatment of convection terms to avoid the numerical diffusion in time (see [3]). In summary, the discretization of equation (7) at point x_j is given by

$$\begin{aligned}
&\frac{11u_j^n - 18u_j^{n-1} + 9u_j^{n-2} - 2u_j^{n-3}}{6\Delta t} \\
&+ u_{j-1/2}^{n-1}\,\frac{4u_j^{n-1} - 5u_{j-1}^{n-1} + u_{j-2}^{n-1}}{3\Delta x} \qquad if\ \ u_{j-1/2}^{n-1} > 0 \\
&- u_{j+1/2}^{n-1}\,\frac{4u_j^{n-1} - 5u_{j+1}^{n-1} + u_{j+2}^{n-1}}{3\Delta x} \qquad if\ \ u_{j+1/2}^{n-1} < 0 \\
&- \nu\,\frac{u_{j+1}^n - 2u_j^n + u_{j-1}^n}{\Delta x^2} = 0
\end{aligned}$$

where $\Delta t, \Delta x$ are the time and space steps and $u_{j-1/2}^{n-1}$, $u_{j+1/2}^{n-1}$ are the midvalues on the left and right cells. So, at each time step we have to solve a well-conditioned symmetric linear system. For Navier-Stokes equations, we use staggered grids for the pressure and the velocity (see figure 3), and the whole

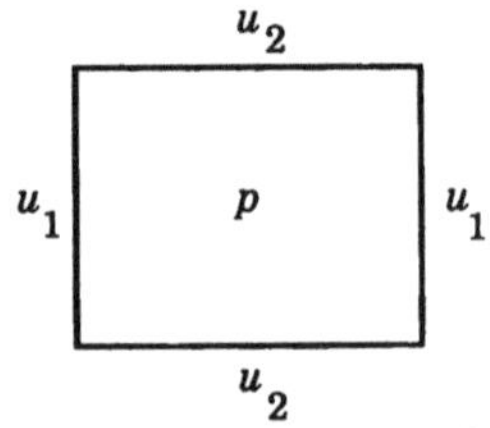

Figure 3. Unknowns on a staggered grid in 2D.

system (1) - (2) is discretized by centered finite differences for the continuity equation at the pressure points, and in the same way than (7) for the two components of the momentum equation at the velocity points. Then we use a cell-by-cell relaxation procedure to solve the discrete system. That means that we solve the whole system (1) - (2) on one cell for the five unknowns of the cell and then go to the next cell. This way, the pressure is updated once as it appears only in the interior of one cell and the velocity is updated twice at each iteration of relaxation. On a given grid, it is well-known that such a relaxation procedure can capture the high frequencies related to the grid in a few iterations but requires a lot of iterations to get the low frequencies. To avoid this problem we use a multigrid method that provides the necessary efficiency. Indeed, we need only two iterations when going down and one iteration when going up in practise on each grid to reach a very good convergence rate. Moreover, at each time step, four V-cycles of the multigrid method are enough to solve the equations. So, if we want to solve a nonlinear problem $Nu = f$ discretized on a grid h by $N_h\, u_h = f_h$, we can write the multigrid algorithm on two consecutive grids as

$$\begin{aligned}
\textit{For } m=1 \textit{ to } M \textit{ do}:\ & \tilde{u}_h^m = \mathcal{R}_h^{J_1}(u_h^{m-1}, f_h) \\
& r_h^m = f_h - N_h\, \tilde{u}_h^m \\
& u_{2h}^m = \mathcal{R}_{2h}^{J_1}(u_{2h}^{m-1}, P_{2h}^h\, r_h^m + N_{2h}\, P_{2h}^h\, \tilde{u}_h^m) \\
& \tilde{\tilde{u}}_h^m = \tilde{u}_h^m + E_h^{2h}(u_{2h}^m - P_{2h}^h\, \tilde{u}_h^m) \\
& u_h^m = \mathcal{R}_h^{J_2}(\tilde{\tilde{u}}_h^m, f_h)
\end{aligned}$$

where $\mathcal{R}$ is the relaxation smoother, P_{2h}^h is the projection operator from grid h to grid $2h$, E_h^{2h} is the extension operator from grid $2h$ to grid h , J_1 and J_2 are the number of iterations of relaxation procedure and M is the number of V-cycles. In practise, we take $J_1 = 2$, $J_2 = 1$ and $M = 4$ for Navier-Stokes problem, at each time step, on a various number of grids up to 8 from a 4×4 coarsest grid to a 512×512 finest grid on the unit square for the driven cavity problem. Besides, as we use staggered grids, both operators P_{2h}^h and E_h^{2h} are founded on interpolation formulas that are chosen of first order.

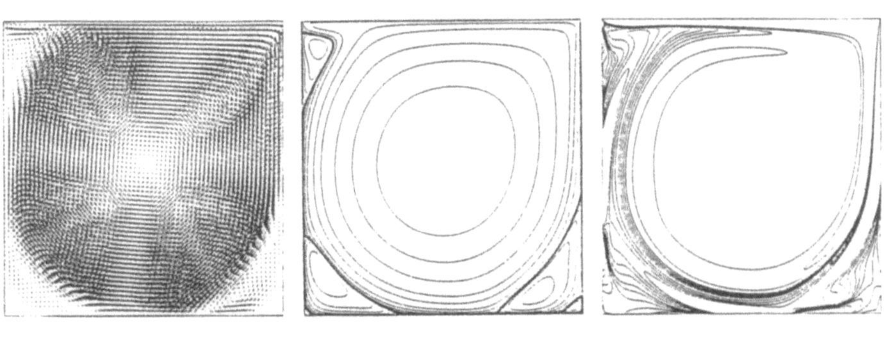

Figure 4. Velocity field, tabulated streamlines and vorticity for the driven cavity problem at Re = 5000.

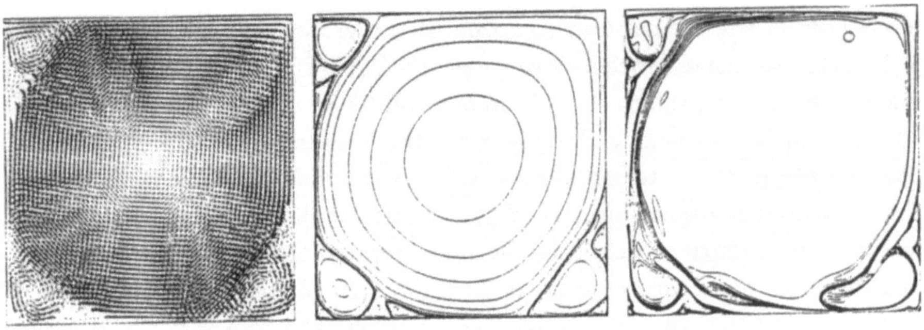

Figure 5. Solution in the driven cavity at Re = 30 000.

4 - Numerical results

We first discuss the results in the driven cavity. At Re = 5000, we obtain a good stable steady solution with a fine representation of secondary and tertiary vortices as it is shown on figure 4. Then, we have found in [4] that the Hopf bifurcation occurs between Re = 5000 and Re = 10000. That is the main difficulty of this problem because the transition to turbulence starts for high Reynolds numbers. So, it is not easy to make a lot of numerical simulations as each of them requires many points in space. Indeed, for such Reynolds numbers, we know that the thickness of the boundary layer is very thin, namely of the order of 1/1000. Besides, the main period of unsteady flows is quite high and then a long time behaviour is needed. On figure 5, we show an unsteady solution for Re = 30000 ; one can see clearly that small eddies develop along the walls and destroy the secondary vortices. The phase portrait at a given point indicates a quasi periodic flow with one main frequency. For this problem, making a complete study of the transition to turbulence is quite out of reach as we have pointed out a tremendous computing power is needed. This remark drives us to the second numerical test concerning the flow behind a cylinder in a channel with solid walls. Here, we have a stable steady symmetric solution for Re = 100 (figure 6) but for Re = 200 the steady solution loses already its stability to the benefit of an unsymmetric purely periodic solution with frequency close to one as it is

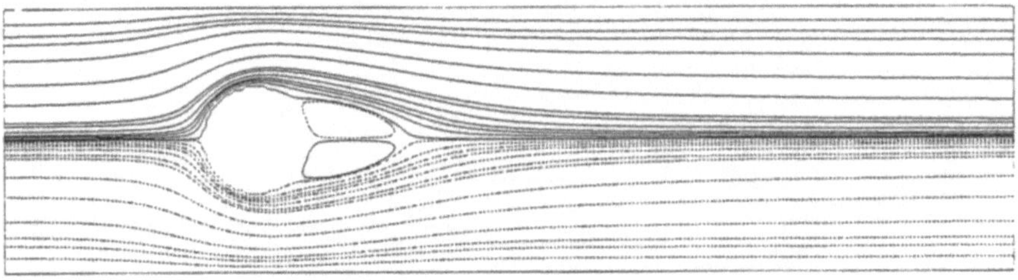

Figure 6. Streamlines in the channel at Re = 100.

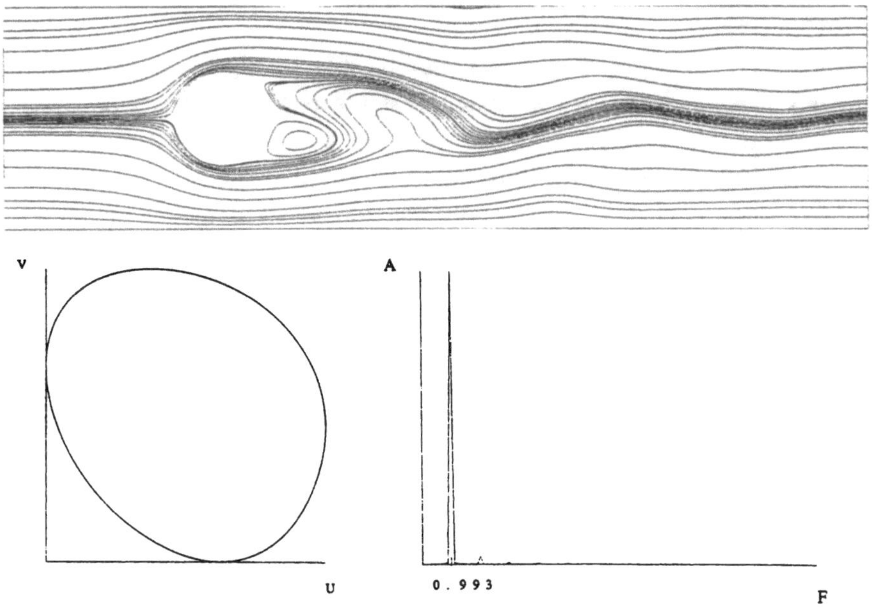

Figure 7. Solution in the channel at Re = 200.

shown on figure 7. For higher Reynolds numbers, the solution becomes more complex and strong vortices are convected downstream. In this case, for *Re* = 1000, we prove the efficiency of boundary condition (6) that allows the vortices to cross the artificial boundary without any reflections (figure 8). When increasing again the Reynolds number, we observe multiple periodic solutions and then chaotic solutions with stretched structures.

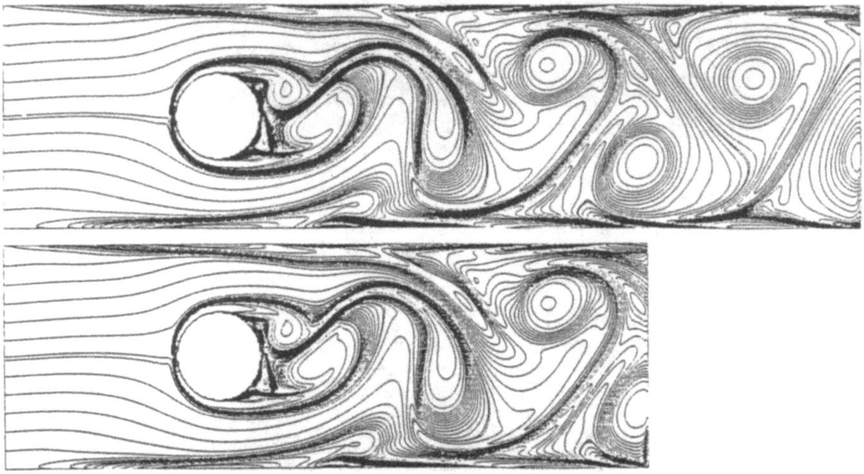

Figure 8. Vorticity in the channel at Re = 1000
for domains of length L = 4 and L = 3.

5 - Conclusions

The method presented in this paper is able to capture efficiently and accurately complex solutions of incompressible Navier-Stokes equations in various geometries. It appears to be a precious tool for numerical simulation of the transition to turbulence that is still under investigation. Moreover, we have tested new open boundary conditions that avoid reflections even when strong vortices are travelling downstream.

Finally, let us point out that the discretization has been carefully handled and that the computed solution has been compared successfully to the exact solution for several 1D problems. So, we believe the solutions of Navier-Stokes equations are quantitatively good although it is not possible to affirm it for sure.

REFERENCES

1. P.M. Gresho, Annu. Rev. Fluid Mech. 23, 413-453 (1991).
2. Ch.H. Bruneau and P. Fabrie, Int. J. Numer. Meth. Fluids, submitted.
3. Ch.H. Bruneau, in Proceeding of the 5th ISCFD, 94-100 (1993).
4. Ch.H. Bruneau and C. Jouron, J. Comp. Phys. 89 n°2, 389-413 (1990).
5. Ch.H. Bruneau, Thèse d'état Université Paris-Sud, (1989).

2nd Int. Coll. on Numerical Analysis, pp. 45-54
D. Bainov and V. Covachev (Eds)

Some applications of the Euclidean algorithm

ADHEMAR BULTHEEL AND MARC VAN BAREL
K.U. Leuven, Dept. Computing Science, Celestijnenlaan 200A, B-3001 Leuven (Belgium)

Abstract We describe some applications of the Euclidean algorithm in modern computational problems like Padé approximation, iterative solution of linear systems, eigenvalue problems, orthogonal polynomials, coding theory etc.

Keywords: Euclidean algorithm, Lanczos method, linear algebra, approximation.

1 EUCLIDEAN ALGORITHM

Euclidean domain: The Euclidean Algorithm (EA) can be traced back to the theorems 1 and 2 of book 7 of the Elements of Euclid, written ca 300 BC, but it is still today of great practical importance in many modern applications, as we want to illustrate in this paper. It computes a GCD of $r, s \in D$, a Euclidean domain as follows. Set $r_{-1} = s$ and $r_0 = r$, and compute $r_k = r_{k-2} \bmod r_{k-1}$ until $r_k = 0$. Then $r_{k-1} = GCD(s, r)$. At every step of the EA we can introduce units (invertible elements from D) x_k and c_k and use $GCD(s_k, r_k) = GCD(s_k x_k, r_k c_k)$ to meet some normalization condition.

EEA and continued fractions: Denote $s_k = r_{k-1}$, then one step of the EA corresponds to the relation $s_{k-1}/r_{k-1} = q_k + r_k/s_k$, or, introducing the units x_k and c_k, we have

$$[s_k \; r_k] = [s_{k-1} \; r_{k-1}]V_k; \quad V_k = \begin{bmatrix} 0 & 1 \\ 1 & -q_k \end{bmatrix} \begin{bmatrix} x_k & 0 \\ 0 & c_k \end{bmatrix} \equiv \begin{bmatrix} y_k & c_k \\ x_k & a_k \end{bmatrix}$$

where $a_k = -q_k c_k$, $y_k = 0$ and $q_k =$ quotient of s_{k-1}/r_{k-1}. Introducing some V_0 with $x_0 = c_0 = 0$ and y_0 and a_0 some units to normalize the given elements s and r as $s_0 = s y_0$ and $r_0 = r a_0$, we arrive after a small manipulation at a continued fraction corresponding to the expansion

$$-\frac{r}{s} = \frac{y_0}{a_0}\left(\frac{c_1|}{|a_1} + x_1\frac{c_2|}{|a_2} + \cdots + x_{k-1}\frac{c_k|}{|a_k} + x_k\frac{r_k|}{|s_k}\right).$$

Truncating this after k terms, we shall get a rational approximation for the left hand side of the form c_{0k}/a_{0k}. The recurrence for the numerators and denominators of the approximants as well as for the tails (residuals) is

$$G_k = G_{-1}\underbrace{V_0 V_1 \ldots V_k}_{V_{0k}} = \begin{bmatrix} y_{0k} & c_{0k} \\ x_{0k} & a_{0k} \\ s_k & r_k \end{bmatrix}; \quad G_{-1} = \begin{bmatrix} 1 & 0 \\ 0 & 1 \\ s & r \end{bmatrix}.$$

This extended recurrence is known as the Extended EA (EEA) [1, 37, 9].

Rational approximation: If we are interested in the rational approximations, then it doesn't matter whether the algorithm will end or not. One of the first applications of this idea is the approximation of a real number by rational numbers. When $r, s \in \mathbb{R}$, then the EEA as described above, taking integer quotients, will generate a sequence of rational numbers approximating $-r/s$. An interesting geometric interpretation of the number theoretic problem can be found in [11] which also includes the multi-dimensional generalization of this problem and the relation with the solution of Diophantine equations.

A similar extension can be imagined for the set of formal Laurent series $D = \mathbb{F}_-\langle z\rangle$ over a field $\mathbb{F}$ having only finitely many terms with positive exponents and where the quotient is the polynomial part of the ratio. Mutatis mutandis, this can be adapted for the formal Laurent series $\mathbb{F}_+\langle z\rangle$ with only finitely many negative exponents.

Complexity, reliability and stability: The computational complexity of the EEA is well known for example for polynomials of degree n, it requires $O(n^2)$ operations (multiplications, additions) to find their GCD. However, it is explained in detail at several places (e.g., [9]) how this can be reduced to $O(n \log_2^2 n)$ when an appropriate divide an conquer technique is applied in conjunction with FFT. On special architectures, the speed can be increased even further. For example, a systolic implementation is discussed in [10].

When the EEA is applied to e.g. Laurent series $\mathbb{F}_-\langle z\rangle$, the successive quotients are of degree 1 or larger. When they all have degree 1, we call it a generic situation. The EEA deals quite naturally with the non-generic situation as well. In many applications however, where the connection with the EEA is not that obvious, great effort has been put in the design of algorithms to deal with breakdown or near-breakdown of the algorithm for a non-generic situation. The point we want to make here is that once the connection with the EEA is made, the adaptation of the algorithm to the non-generic situation, avoiding exact breakdown (reliability) and near-breakdown (stability) is rather obvious.

2 APPROXIMATION OF FORMAL SERIES

One possible application is the approximation of formal Laurent series. Suppose we work in $\mathbb{F}_-\langle z\rangle$ as explained in section 1. The successive rational approximants for $f = -r/s$ (suppose for simplicity that $s = -1$, so that $f = r$) satisfy $f - c_{0k}/a_{0k} = r_k/a_{0k}$. When the "quotients" have $\deg a_k(z) = \alpha_k$, then it can be shown that for a strictly proper f: $\deg c_{0k} < \deg a_{0k} = \alpha_{0k} = \alpha_1 + \cdots + \alpha_k$ and $\deg(r_k/a_{0k}) = -2\alpha_{0k} - \alpha_{k+1}$ with $\alpha_{k+1} \geq 1$. Comparing the degrees of freedom in the approximant ($2\alpha_{0k}$) and the number of interpolation conditions that are met, we see that we have solved here a Padé approximation problem at infinity [4]. When f is the transfer function of a linear system, then the approximant is the one of minimal (McMillan) degree which fits the first $2\alpha_{0k}$ coefficients. This is known as a solution of the minimal partial realization (MPR) problem [36]. In this context, the EEA is closely related to the Berlekamp-Massey algorithm, which was first designed for shift register synthesis of linear predictive codes. The numbers α_{0k} are known as Kronecker indices. The numbers α_k did not seem to have received a standard name. We shall call them Euclidean indices.

Similarly, the rational approximants produced by the EEA for the ratio of two elements from $\mathbb{F}_+\langle z\rangle$ are convergents of a continued fraction which is known as a P-fraction, a name due to Arne Magnus [39]. The P stands for principal part (plus constant term) which is indeed what is meant by the quotient here. The approximants are known as Padé approximants (at the origin, but this is what is usually meant by Padé approximation). One of the earlier papers to explicitly recognize the EA as a method for computing Padé approximants is [42].

In the literature of Padé approximation, the notion of Padé table is a convenient concept to formulate certain results. The table is a two dimensional array which contains in the entry (i,j) a Padé approximant with numerator degree i and denominator degree j. In general the Padé table consists of square blocks of size 1 or larger. The entries in one block are all equal. A block of size > 1 is called singular. A Padé table with singular blocks is called non-normal. The EA computes Padé approximants on diagonals of the table thereby jumping from block to block.

3 LINEAR ALGEBRA AND ORTHOGONAL POLYNOMIALS

The interpolation conditions for the MPR problem in a linearized form can be written as a Hankel system of equations. Suppose we denote by bold face letters the vertical matrices containing the coefficients of series or polynomials, then the condition that for $f(z) \in \mathbb{F}_-\langle z \rangle$, $\deg a_{0k}(z) \leq \alpha_{0k}$ there should exist a polynomial $c_{0k}(z)$ of degree $< \alpha_{0k}$ such that $\deg r_k < -\alpha_{0k}$ with $r_k(z) = f(z)a_{0k}(z) - c_{0k}(z)$, can be written as $H\mathbf{a}_{0k} = \mathbf{r}_k$ with $H = H(f)$ the Hankel matrix with symbol f. With the down shift matrix Z we can define blocks $A_{.k} = [\mathbf{a}_{0k}|Z\mathbf{a}_{0k}|\ldots|Z^{\alpha_{k+1}-1}\mathbf{a}_{0k}]$ and the upper triangular block matrix $A = [A_{.0}|A_{.1}|\ldots]$. It turns out that $HA = R$ (R block lower triangular) and $A^THA = D$ with D block diagonal.

This relation can be interpreted in two different ways: (1) as a block triangular factorization of H^{-1} and, (2) as a block orthogonality relation for the polynomials a_{0k} with respect to the moment matrix H. By block orthogonality we mean that a polynomial in a block is orthogonal to all the polynomials in previous blocks, but it may not be orthogonal to all the polynomials in the same block (See e.g. [15]). The matrix H is not necessarily positive definite or strongly regular (some of its leading submatrices may be singular). In fact the sizes of the blocks in the decomposition correspond to the number of consecutive leading submatrices that are singular. In terms of Padé approximation, this corresponds to a part of the diagonal path passing through a singular block of the Padé table.

The orthogonal polynomials satisfy the recurrence relation $a_{0k}(z) = a_{0,k-2}(z)x_{k-1}c_k + a_{0,k-1}(z)a_k(z)$ which follows immediately from the EEA. As before, the x_{k-1} and c_k are units and $a_k(z) = -q_k(z)c_k$ is the "quotient" of $s_{k-1}(z)$ and $r_{k-1}(z)$. It represents a generalization of the 3-term recurrence relation for classical orthogonal polynomials. This recurrence relation can again be written as the matrix relation $ZA = AT$, with T a block tridiagonal which is also a sparse upper Hessenberg matrix. It generalizes the Jacobi matrix from the theory of orthogonal polynomials. An equivalent formulation of the previous relation is $A^TH(zf)A = DT = J$ with J a symmetric block tridiagonal. For a discussion of these matters see for example [22, 31, 33, 16].

4 ITERATIVE SOLUTION OF LINEAR SYSTEMS OF EQUATIONS

During the last five years an enormous effort has been done to develop and apply the previous theory in the context of the iterative solution of linear systems with so called Krylov subspace methods. The Lanczos algorithm has long been believed to be only numerically acceptable for Hermitean positive definite systems because of the numerical difficulties one may expect when the algorithm is applied in a non-generic situation when some numbers become zero or nearly zero. However, as in previous situations, the EEA may also here bring a solution to overcome such a breakdown whenever possible (curable breakdown).

Suppose we want to solve the system $Mx = b$. The size of the matrix M is m. The residual is $r = b - Mx$. For some starting vector x_0, successive approximants x_k are obtained by a conjugate direction method minimizing the residual norm. It turns out that the residuals $r_k = b - Mx_k$ are elements from the Krylov subspaces $\mathcal{K}_k(r_0, M) = \text{span}\{r_0, Mr_0, \ldots, M^k r_0\}$. More precisely, there exists a polynomial $\varphi_k(z)$ of degree k and $\varphi_k(0) = 1$ such that $r_k = \varphi_k(M)r_0$. Let us first suppose that M is symmetric. Then minimizing $\|r_k\|^2 = r_k^T r_k$ is equivalent to minimizing $\|\varphi_k\|_H^2$ where the latter norm is taken with respect to the Hankel moment matrix $H = H(f)$ where $f(z) = \sum f_k z^{-k}$ and $f_k = r_0^T M^{k-1} r_0$. The comonic polynomial of minimal norm is an orthogonal one, hence we get the orthogonality relations $\varphi_k \perp \varphi_l$ for $k \neq l$ with respect to the metric H or equivalently $r_k \perp r_l$ in the metric I. Thus in principle we can perform the iterations of the Lanczos algorithm whenever we can construct the orthogonal polynomials φ_k.

The problem of breakdown occurs when the Hankel matrix $H = H(f)$ is not strongly regular, which can happen when M is not positive definite. But we are then in a situation where we have to construct formal orthogonal polynomials with respect to an indefinite Hankel moment matrix, and this problem has been treated in the previous section.

The previous formulas can be translated into the present terminology as follows. First we consider for $n < m$ the matrix $Y_n = [y_0|y_1|\ldots|y_n]$ where $y_0 = r_0$ and $y_k = My_{k-1}$ for $n \geq k \geq 1$. Then, by definition, the leading submatrix of size $n+1$ of H is given by $H_n = Y_n^T Y_n$. The Krylov space is the column space of Y_n and the problem is to construct an orthogonal basis for this space. We have to find an upper triangular matrix A_n such that the columns of $\hat{Y}_n = Y_n A_n$ are orthogonal vectors, or in the presence of singular blocks, they should be block orthogonal vectors. The matrix A_n is a submatrix of the matrix A we had in the previous section and which can be constructed recursively from the data f or equivalently from the matrix Y_n by the EEA. The orthogonality means that $\hat{Y}_n^T \hat{Y}_n = A_n^T H_n A_n = D_n$ with D_n block diagonal. Moreover we have that $\hat{Y}_n M \hat{Y}_n = A_n^T H_n(zf) A_n = D_n T_n = J_n$ is a block tridiagonal matrix. Note that in the block case, the index n refers to the number of blocks, rather than the number of columns.

At first sight, the case of a non symmetric matrix M, can not be treated by the EA because the matrix with entries $r_0(M^T)^i M^j r_0$ is not Hankel anymore. One possibility is to construct orthogonal polynomials for a general moment matrix, but this is computationally less efficient because the Jacobi matrix will then not be tridiagonal or block tridiagonal, but it will be a full upper Hessenberg matrix. Algorithms of this type belong to the class of Arnoldi-like algorithms. However, it is still possible, with substantially less computational effort to generate a Hankel matrix as before. Indeed, we consider two Krylov subspaces: the one generated by the vectors y_i and another one $\mathcal{K}_n(r_0, M^T)$, which is spanned by the columns of $Z_n = [z_0|z_1|\ldots|z_n]$ with $z_0 = r_0$ and $z_k = M^T z_{k-1}$. Clearly, the matrix $H_n = H_n(f) = Z_n^T Y_n$ is again Hankel with symbol $f(z) = z_0^T (zI - M)^{-1} y_0$. Applying the EA as before to get the upper triangular factor A_n, leads to the fact that $\hat{Z}_n = Z_n A_n$ and $\hat{Y}_n = Y_n A_n$ satisfy a block biorthogonality condition $\hat{Z}_n^T \hat{Y}_n = A_n^T H_n A_n = D_n$. For a survey of recent developments see [26, 33, 16].

5 EIGENVALUE PROBLEM AND LINEAR SYSTEMS

Let us consider a linear system which can be described by the state space equations

$$\begin{aligned} x_{k+1} &= Ax_k + Bu_k, & A &\in \mathbb{F}^{n\times n}, B \in \mathbb{F}^{n\times 1} \\ y_{k+1} &= C^T x_{k+1}, & C &\in \mathbb{F}^{n\times 1} \end{aligned}$$

The vector x_k is the state, u_k is the input and y_k is the output. The triple (A, B, C) is called a state space realization of the system. When the number n (dimension of the state space) is the smallest possible one to describe the input-output behaviour of the system, the realization is called minimal.

After taking z-transforms (e.g., $u(z) = \sum u_k z^{-k}$), we obtain $y(z) = f(z)u(z)$ with transfer function $f(z) = C^T(zI - A)^{-1}B$. The coefficients in its expansion at ∞, i.e., $f_k = C^T A^{k-1} B$ are called Markov parameters, and the sequence (f_k) is called the impulse response. Thus the MPR problem is to match a certain number of Markov parameters by constructing a rational transfer function of the smallest possible degree. It is a Padé approximation problem at infinity for the transfer function of the system.

The matrix $\mathcal{O} = [C\ A^TC\ (A^T)^2C\ \ldots]^T$ is called the observability matrix and $\mathcal{R} = [B\ AB\ A^2B\ \ldots]$ is called the reachability matrix. The system is called reachable when $\mathcal{R}$ has full rank, i.e., $n = \operatorname{rank}\mathcal{R}$ and it is called observable when $\mathcal{O}$ has full rank, i.e., when $n = \operatorname{rank}\mathcal{O}$. When $n = \operatorname{rank}\mathcal{R} = \operatorname{rank}\mathcal{O} = \operatorname{rank}\mathcal{OR}$, then the realization is minimal. Note that $H = \mathcal{OR}$ is the same as the Hankel matrix $H(f)$, hence $\operatorname{rank} H(f) = \deg f$, if f is reduced.

To see that the Lanczos algorithm can also help in reducing the eigenvalue problem for the matrix M to a simpler one, suppose that n refers to the block index such that D_n is regular. Then note that from (notation of previous section) $\hat{Z}_n^T\hat{Y}_n = D_n$ and hence $\hat{Z}_n^T = D_n\hat{Y}_n^{-1}$, it follows from $\hat{Z}_n^T M\hat{Y}_n = D_nT_n$ that $\hat{Y}_n^{-1}M\hat{Y}_n = T_n$. This means that M is similar to T_n and thus it has the same eigenvalues. Since T_n is much simpler, being a sparse block tridiagonal, its eigenvalues are easily obtained. Since the finite dimensional analog of $ZA = AT$ reads $F(a_{0,n+1})A_n = A_nT_n$, where $F(a_{0,n+1})$ represents the Frobenius matrix for the polynomial $a_{0,n+1}(z)$, also $T_n = A_n^{-1}F(a_{0,n+1})A_n$ and thus it follows that the eigenvalues of M are eigenvalues of T_n and therefore zeros of $a_{0,n+1}(z)$.

All this is true when there is some $\alpha_{0,n+1}$ which is equal to m, the size of the matrix M. This need not be true in general. First note that the input for the EA is essentially $f(z) = z_0^T(zI - M)^{-1}y_0$. The triple (M, y_0, z_0) is a state space realization for f. But if it is not minimal, then M may contain a larger spectrum than can be obtained as poles from f. The EA or Lanczos algorithm will end because $\alpha_{0,n+1} = r = \deg f < m$. It will thus be impossible to recover the whole spectrum of M from f. The Lanczos algorithm will have a breakdown, which can not be cured by the principles of the EA. Note that nonminimality is no problem in the solution of systems, since this breakdown indicates that f is approximated exactly and the residual will be zero. For the eigenvalue problem however, it turns out that there are two possibilities to be considered. Define $r_y = \operatorname{rank} Y_\infty = \dim\mathcal{K}_\infty(y_0, M)$ and $r_z = \operatorname{rank} Z_\infty = \dim\mathcal{K}_\infty(z_0, M^T)$, then either (A) or (B) will happen.

(A) $m > r = \min\{r_y, r_z\}$, then the system is reachable or observable, and we can recover the spectrum of M restricted to the reachability space (that is $\mathcal{K}_\infty(z_0, M^T) =$ the column space of Z_∞) or the observability space (that is $\mathcal{K}_\infty(y_0, M) =$ the column space of Y_∞).

(B) $r < \min\{r_y, r_z\}$, then the starting vectors y_0 and/or z_0 were chosen inadequately so that they do not have components along a basis for the maximal Jordan block that can be associated with each eigenvalue of M. The algorithm is trapped in a smaller dimensional space and only the spectral information of M restricted to these smaller spaces will be recovered. See [33, 44, 16].

6 CODING THEORY

Another classical application of the Euclidean algorithm can be found in the theory of error correcting codes, more precisely Goppa codes. Two examples are BCH codes and Reed-Solomon codes. A comprehensive treatment of the subject can be found in [41]. The Berlekamp and Massey algorithm (BMA) [8, 40], was first developed in this context to find linear feedback shift register realizations for the associated system. In a sense, this is equivalent to the Euclidean algorithm. It requires inner products though, hence it is a layer adjoining variant [13] of the EA and therefore less suited for parallel implementation. See also [23, 24] for generalizations of the BMA to multi-sequences.

7 MATRIX AND OTHER GENERALIZATIONS

Most of the previous notions concerning the EA can be generalized to the $q \times p$ dimensional situation. There are several possibilities for generalizing all the scalar notions and what might be understood by a EA for this situation is not uniquely defined.

Linear systems: The theory of linear systems was probably the first place where matrix generalizations of the previous theory were developed. The concept of state space is remarkably easy to adapt to systems with multiple inputs and multiple outputs. In the realization triple (A, B, C) one should assume that $B \in \mathbb{F}^{n\times p}$ and $C \in \mathbb{F}^{n\times q}$ where p is the dimension of the input vector u_k and q is the dimension of the output vector y_k. The transfer function becomes a matrix rational function of size $q \times p$. There are several possibilities to write a rational matrix function as the "ratio" of two polynomial matrices. For example $f(z) = s(z)^{-1}r(z)$ with s and r polynomial $r \in \mathbb{F}^{q\times p}[z]$ and $s \in \mathbb{F}^{q\times q}[z]$ is a left description. The Hankel matrix $H(f)$ is replaced by a block Hankel matrix with $q \times p$ blocks.

Greatest common divisor: The definition of greatest common left divisor (GCLD) for matrix polynomials can be found in [36, p.376-378]. You can also find there a theorem saying that g is a GCLD of s and r when there exists a unimodular matrix V such that $[s\ r]V = [g\ 0]$. A unimodular matrix is a polynomial matrix whose inverse is again a polynomial matrix. Thus the matrix version of the EEA will, exactly like in the scalar case, start from G_{-1} and generate successively $G_k = G_{k-1}V_k$ until after a finite number of steps one has $[s_n\ r_n]$ with $r_n = 0$. The matrices V_k are a composition of elementary Gaussian-type elimination matrices involving powers of z off the diagonal. Their description can be found in [48] for the MPR problem and in [14] for the matrix Padé problem. On a more theoretical basis, the same ideas are found in [3]. Predecessors are [21] and [2].

Approximation of series: When the previous algorithm is used for matrix polynomials s and r, then it will end after a finite number of steps. In general it may be applied to matrix Laurent series and then it may never stop, but at every stage, the approximants $c_{0k}a_{0k}^{-1}$ are matrix versions of the Padé approximants at ∞ or at 0 of the rational matrix function $f = -s^{-1}r$. Again we refer to the papers mentioned above for the details. See also [29].

Matrix continued fractions: We can also associate with the matrices V_k an elementary step of a matrix continued fraction. However, in the scalar case, the left top element y_k was zero. In the matrix case, this will no longer hold. This implies that the matrix generalization of the continued fraction will not only continue the division steps in the denominator, but it will also show a continuing division in the numerator. More precisely, the convergents of the continued fraction on the right in the expansion

$$-s^{-1}r = \left[y_0\left(c_1 + y_1\frac{c_2 + y_2\frac{\cdots}{\cdots}}{a_2 + x_2\frac{\cdots}{\cdots}}\right)\right]\left[a_0\left(a_1 + x_1\frac{c_2 + y_2\frac{\cdots}{\cdots}}{a_2 + x_2\frac{\cdots}{\cdots}}\right)\right]^{-1}; \quad \text{where } \frac{a}{b} = ab^{-1}$$

are the successive approximants.

Multi-point interpolation: Until now, we only mentioned rational approximation with interpolation conditions at 0 or ∞. Of course, it would not be difficult to replace the indeterminate z by $x - \alpha$ which would result in a Laurent series in the neighborhood of α, or one could consider simultaneous approximation of two series: one at 0 and one at ∞ or even at more points, giving multi-point rational approximations. The technicalities are of course more involved, but basically the same. Whether the reader is still prepared to call this a variant of the Euclidean algorithm is of course a matter of taste.

By now multipoint and matrix versions have been considered [51, 5, 7, 50, 49]. The block structure of some these problems has been discussed recently [32, 6].

8 DUALITY

As en example of the previous generalizations, we mention the vector valued Padé approximation problem, which has a longer history. We use this context to bring a remarkable duality to the foreground. There are two types of problems that were already considered by Hermite.

There is the problem of finding polynomials $p_0, \ldots, p_n$ with certain bounds $\delta_0, \ldots, \delta_n$ on their degrees such that they approximate given power series $f_0, \ldots, f_n$ up to a certain order in the sense that $\sum_0^n p_k(z) f_k(z) = O(z^\omega)$ ($\omega = \sum_k \delta_k + n$ at least). This is known as the Padé-Hermite approximation problem.

On the other hand, there is the problem of simultaneous Padé approximation where polynomials $q_0, \ldots, q_n$ are to be found such that for given series $f_1, \ldots, f_n$ and given integer ω, it holds that $f_k(z) q_0(z) - f_0(z) q_k(z) = O(z^\omega)$ and $\deg q_k \leq \omega - \delta_k$.

There is an intimate relation between the two problems. Solving one problem is equivalent to solving the other. For a discussion of the general duality principle in vector Padé-Hermite problems, see [19, 38].

9 LOOK-AHEAD STRATEGIES

The latest successes that were obtained in this circle of problems is a breakthrough in improving the numerical stability of this type of algorithms. This is the problem we mentioned before of how to make the algorithms deal with near-breakdown, and not just with exact breakdown.

Taylor was probably the first to propose a look-ahead strategy in the context of the Lanczos algorithm, see [47, 45]. These papers coined the term look-ahead. See also [12, 25, 27, 28]. The same idea has been implemented for the related problems that were discussed in this survey, like for example Padé approximation [32], the solution of Hankel [43, 18] and Toeplitz systems [54, 35, 34] and in general rational (matrix) interpolation problems [52].

The EA gives a reliable algorithm because it jumps over the singular blocks in the Padé table, or in terms of Hankel matrices, it skips the singular leading submatrices. Such a singularity is detected by a pivot element being zero. Near-breakdown is detected when this pivot is small resulting in a dramatic loss of accuracy. Thus numerically, it makes more sense not to leap from a nonsingular situation to the next one, but to leap from a well conditioned situation to a well conditioned one. This then results in a stable EEA (SEEA).

Because a well-conditioned situation is a nonsingular one by definition, the transition matrix V_k^s we shall need in a SEEA to jump from one approximant to the next one, will be a product of some of the V_k matrices of the EEA. Therefore, the matrix V_k^s will not have the same simple form as a V_k with $y_k = 0$, x_k and c_k units and only a_k polynomial. In general V_k^s will have polynomial entries in all 4 places, but it will still be unimodular.

The same idea of grouping together several steps of the simple EEA was already used in the superfast versions of the EEA. Suppose we need N steps of a classical EEA to reach a solution, then the divide and conquer strategy grouped the product $V_1 \dots V_N$ as $V_{1n}V_{n+1,N}$ where $n = \lfloor N/2 \rfloor$. In terms of Padé approximation, this means that to have an approximant of degree $\alpha_{0,N+1}$ of $f = -s^{-1}r$, we may compute a degree $\alpha_{0,n+1}$ approximant for f and a degree $\alpha_{0,N+1} - \alpha_{0,n+1}$ approximant for the tail of the continued fraction. These two can then be combined to give the desired approximant. By applying this strategy recursively, the superfast algorithms emerge.

In the previous discussion, a leap from a well conditioned point to a well conditioned point was always supposed to be on a path that is normally followed by the EEA, thus on a diagonal path. By introducing shift parameters [50, 49] this extends to any parallel diagonal. By changing the number of interpolation conditions as well as the degree conditions, one can walk in principle from any sonsingular (well conditioned) point to any other nonsingular (well conditioned) point in the interpolation table. The idea is that the solution to an interpolation problem can be written as a system of homogeneous linear equations. The space of all the solutions for this problem is a submodule which is generated as polynomial multiples of some basis of polynomial vectors.

10 CONCLUDING REMARKS

In these applications, the farther we are away from the simple scalar case, the more difficult it will be to recognise the Euclidean algorithm. If one is prepared to push this a little further, one could design Lanczos type algorithms for Toeplitz matrices or even general matrices [17] or algorithms for quasi-Hankel or Toeplitz matrices or structured matrices in general (e.g., [46, 38]).

Another important challenge for future research is to see how time-variant system theory can be applied to the circle of ideas we have proposed here. In time-variant systems, the matrices A, B and C also depend on k. See [30, 53, 20].

Acknowledgement: This research was supported by the National Fund for Scientific Research (NFWO), project Lanczos, grant #2.0042.93.

References

[1] A.V. Aho, J.E. Hopcroft, and J.D. Ullman. *The design and analysis of computer algorithms.* Addison Wesley, Reading, Mass., 1974.

[2] B.M. Anderson, F.M. Brasch, and P.V. Lopresti. The sequential construction of minimal partial realizations from finite input-output data. *SIAM J. Control*, 13(3):552–571, May 1975.

[3] A.C. Antoulas. On recursiveness and related topics in linear systems. *IEEE Trans Automatic Control*, AC-31:1121–1134, 1986.

[4] G.A. Baker, Jr. and P.R. Graves-Morris. *Padé Approximants. Part I: Basic Theory*, volume 13 of *Encyclopedia of mathematics and its applications.* Addison-Wesley Publishing Co., Reading, 1981.

[5] B. Beckermann. *Zur Interpolation mit polynomialen Linearkombinationen beliebiger Funktionen.* PhD thesis, Universität Hannover, 1989.

[6] B. Beckermann, J. Gilewicz, and V. Kaliaguine. On the definition and normality of a general table of simultaneous Padé approximants. Technical Report CPT-92/P.2686, CNRS Luminy, France, 1992.

[7] B. Beckermann and G. Labahn. A uniform approach for Hermite-Padé and simultaneous Padé approximants and their matrix-type generalizations. *Numer. Alg.*, 3(1-4):45–54, 1992.

[8] E.R. Berlekamp. *Algebraic coding theory.* McGraw-Hill, New York, 1968.

[9] R.P. Brent, F.G. Gustavson, and D.Y.Y. Yun. Fast solution of Toeplitz systems of equations and computation of Padé approximants. *J. Algorithms*, 1:259–295, 1980.

[10] R.P. Brent and H.T. Kung. Systolic VLSI arrays for polynomial GCD computation. *IEEE Trans. on computers*, C-33(8):731–736, 1984.

[11] A.J. Brentjes. *Multi-dimensional continued fraction algorithms*, volume 145. Mathematical Centre, Amsterdam, The Netherlands, 1981.

[12] C. Brezinski, M. Redivo Zaglia, and H. Sadok. Avoiding breakdown and near-breakdown in Lanczos type algorithms. *Numerical Algorithms*, 1(3):261–284, 1991. Addendum, vol. 2(2):133-136, 1992.

[13] A.M. Bruckstein and T. Kailath. Inverse scattering for discrete transmission-line models. *SIAM Review*, 29(3):359–389, 1987.

[14] A. Bultheel and M. Van Barel. A matrix Euclidean algorithm and the matrix minimal Padé approximation problem. In C. Brezinski, editor, *Continued fractions and Padé approximants*, pages 11–51. North-Holland, 1990.

[15] A. Bultheel and M. Van Barel. Formal orthogonal polynomials and rational approximation for arbitrary bilinear forms. Technical Report TW163, Department of Computer Science, K.U. Leuven, December 1992.

[16] A. Bultheel and M. Van Barel. Euclid, Padé and Lanczos, another golden braid. Technical Report TW188, Department of Computer Science, K.U. Leuven, April 1993.

[17] A. Bultheel and M. Van Barel. Linear algebra, rational approximation and orthogonal polynomials, 1993. (In preparation).

[18] S. Cabay and R. Meleshko. A weakly stable algorithm for Padé approximants and the inversion of Hankel matrices. *SIAM J. Matrix Anal. Appl.*, 14:735–765, 1993.

[19] G. De Samblanx. Vektor Padé-Hermite benaderingen (Vector Padé-Hermite approximation). Master's thesis, K.U. Leuven, Dept. Computing Sci., 1993. (In Dutch)

[20] P. Dewilde, M.A. Kaashoek, and M. Verhaegen, editors. *Challenges of a generalized, non-stationary system theory*. Essays of the Dutch Academy of Arts and Sciences. Dutch Acad. Arts Sci., Amsterdam, 1993.

[21] B.W. Dickinson, M. Morf, and T. Kailath. A minimal realization algorithm for matrix sequences. *IEEE Trans. Autom. Control*, AC-19:31–38, 1974.

[22] A. Draux. *Polynômes orthogonaux formels – applications*, volume 974 of *Lecture Notes Mathematics*. Springer-Verlag, Berlin, 1983.

[23] Gui-Liang Feng and K.K. Tzeng. A generalized Euclidean algorithm for multisequence shift-register synthesis. *IEEE Trans. Information Theory*, 35(3):584–594, 1989.

[24] Gui-Liang Feng and K.K. Tzeng. A generalization of the Berlekamp-Massey algorithm for multisequence shift-register synthesis with applications to decoding cyclic codes. *IEEE Trans. Information Theory*, 37(5):1274–1287, 1991.

[25] R.W. Freund. The look-ahead Lanczos process for large nonsymmetric matrices and related algorithms. In M.S. Moonen, G. Golub, and B.L.R. De Moor, editors, *Linear algebra for large scale and real-time applications*, volume 232 of *NATO-ASI series E: Applied Sciences*, pages 137–163. Kluwer Acad. Publ., Dordrecht, 1993.

[26] R.W. Freund, G.H. Golub, and N.M. Nachtigal. Iterative solution of linear systems. *Acta Numerica*, 1:57–100, 1992.

[27] R.W. Freund, M.H. Gutknecht, and N.M. Nachtigal. An implementation of the look-ahead Lanczos algorithm for non-Hermitian matrices, Part I. *SIAM J. Sci. Stat. Comp.*, 14(1):137–158, 1993.

[28] R.W. Freund and N.M. Nachtigal. An implementation of the look-ahead Lanczos algorithm for non-Hermitian matrices, Part II. Technical Report 90.46, RIACS, NASA Ames Research Center, Moffet Field, 1990.

[29] P.A. Fuhrmann. A matrix Euclidean algorithm and matrix continued fraction expansion. *Systems and Control Letters*, 3:263–271, 1983.

[30] I. Gohberg, editor. *Time-variant systems and interpolation*, volume 56 of *Operator Theory,*

Advances and Applications. Birkhäuser Verlag, 1992.

[31] W.B. Gragg and A. Lindquist. On the partial realization problem. *Lin. Alg. Appl.*, 50:277–319, 1983.

[32] M.H. Gutknecht. Block structure and recursiveness in rational interpolation. In E.W. Cheney, C.K. Chui, and L.L. Schumaker, editors, *Approximation Theory VI*, pages 93–130. Academic Press, 1992.

[33] M.H. Gutknecht. A completed theory of the unsymmetric Lanczos process and related algorithms. Part I. *SIAM J. Matrix Anal. Appl.*, 13(2):594–639, 1992.

[34] M.H. Gutknecht. Stable row recurrences for the Padé table and generically superfast lookahead solvers for non-Hermitian Toeplitz systems. *Lin. Alg. Appl.*, 188/189:351–422, 1993.

[35] M.H. Gutknecht and M. Hochbruck. Look-ahead Levinson and Schur algorithms for non-Hermitian Toeplitz systems. Technical Report 93-11, IPS, ETH Zürich, 1993.

[36] T. Kailath. *Linear Systems*. Prentice-Hall, Inc., Englewood Cliffs, N.J., 1980.

[37] D.E. Knuth. *The art of computer programming, Vol 2 : Seminumerical algorithms*. Addison Wesley, Reading, Mass., 1969.

[38] G. Labahn. Inversion components of block Hankel-like matrices. *Lin. Alg. Appl.*, 177:7–48, 1992.

[39] A. Magnus. Expansion of power series into P-fractions. *Math. Z.*, 80:209–216, 1962.

[40] J.L. Massey. Shift-register synthesis and BCH decoding. *IEEE Trans. Inf. Theory*, IT-15:122–127, 1969.

[41] R.J. McEliece. *The theory of information and coding: A mathematical framework for communication*, volume 3 of *Encyclopedia of mathematics*. Addison Wesley, Reading, Mass., 1977.

[42] R.J. McEliece and J.B. Shearer. A property of Euclid's algorithm and an application to Padé approximation. *SIAM J. Appl. Math.*, 34:611–616, 1978.

[43] R.J. Meleshko. *A stable algorithm for the computation of Padé approximants*. PhD thesis, Dept. of Computing Sc., Univ. of Alberta, Canada, 1990.

[44] B.N. Parlett. Reduction to tridiagonal form and minimal realizations. *SIAM J. Matr. Anal. Appl.*, 13:567–593, 1992.

[45] B.N. Parlett, D.R. Taylor, and Z.A. Liu. A look-ahead Lanczos algorithm for unsymmetric matrices. *Mathematics of Comp.*, 44(169):105–124, 1985.

[46] A.H. Sayed. *Displacement structure in signal processing and mathematics*. PhD thesis, Stanford University, August 1992.

[47] D.R. Taylor. *Analysis of the look ahead Lanczos algorithm*. PhD thesis, Center for Pure and Applied Mathematics, University of California, Berkeley, 1982.

[48] M. Van Barel and A. Bultheel. A matrix Euclidean algorithm for minimal partial realization. *Linear Algebra Appl.*, 121:674–682, 1989.

[49] M. Van Barel and A. Bultheel. A new approach to the rational interpolation problem: the vector case. *J. Comput. Appl. Math.*, 33(3):331–346, 1990.

[50] M. Van Barel and A. Bultheel. A general module theoretic framework for vector M-Padé and matrix rational interpolation. *Numer. Alg.*, 3:451–461, 1992.

[51] M. Van Barel and A. Bultheel. A new formal approach to the rational interpolation problem. *Numer. Math.*, 62:87–122, 1992.

[52] M. Van Barel and A. Bultheel. The "look-ahead" philosophy applied to matrix rational interpolation problems. In *Proceedings MTNS, Regensburg, Germany, August 1993*, 1993. Submitted.

[53] A.-J. van der Veen. *Time-variant system theory and computational modeling. Realization, approximation and factorization*. PhD thesis, Technical University Delft, The Netherlands, June 1993.

[54] P. Vyvey. Oplossen van niet sterk reguliere Toeplitz stelsels (The solution of non strongly regular Toeplitz systems). Master's thesis, K.U. Leuven, Dept. Computing Sci., 1987. (In Dutch).

2nd Int. Coll. on Numerical Analysis, pp. 55-61
D. Bainov and V. Covachev (Eds)

Best Approximation in Reproducing Kernel Hilbert Spaces

DU-WON BYUN and SABUROU SAITOH

Department of Mathematics, Global Analysis Research Center, Seoul National University, Seoul 151-742, Korea

Department of Mathemaics, Faculty of Engineering, Gunma University, Kiryu 376, Japan

Abstract. Let H_K be a reproducing kernel Hilbert space on a set E, and for a subset X of E let $H(X)$ be a Hilbert space composing of functions on X. Then, under some reasonable conditions on H_K and $H(X)$, the best approximation problem in the sense of

$$\inf_{f \in H_K} \|Tf - F\|_{H(X)}$$

is examined for $F \in H(X)$ and for the restriction $Tf = f|_X$ to the set X. The approximation problems by the solutions of the heat equation are examined as one concrete application of the general results.

Keywords: Reproducing kernel, best approximation, heat equation

1. INTRODUCTION

Let E be an arbitrary set, and let H_K be a Hilbert (possibly finite dimensional) space composed of complex - valued functions f on E admitting a reproducing kernel $K(p,q)$.

Meanwhile, for any subset X of E we consider a Hilbert space $H(X)$ comprising of functions F on X. In the relationship of two Hilbert spaces H_K and $H(X)$, we assume that

(a) for the restriction $f|_X$ of the members f of H_K to the set X, $f|_X$ belongs to the Hilbert space $H(X)$,
and

(b) the linear operator $Lf = f|_X$ is continuous from H_K into $H(X)$.

Then, we shall consider the fundamental problem

$$\inf_{f \in H_K} \|Lf - F\|_{H(X)}$$

for a member F of $H(X)$.

For the sake of the nice properties of L and its adjoint L^* in our situation, we will be able to obtain "algorithms " to decide whether the best approximations f^* of F in the sense of

$$\inf_{f \in H_K} \|Lf - F\|_{H(X)} = \|Lf^* - F\|_{H(X)}$$

exist. Furthermore, when there exists the best approximations f^*, we will be able to get constructively the best approximations f^*. Indeed, we shall give intrinsic representations of the best approximations f^* in terms of the given function F.

In this paper, we first establish a general and abstract theorem in the above problem as an original work and, secondly we shall give a typical and concrete example. This example is contained in [7]. Other application to concrete problems and related topics were submitted to the journals as separated papers in [3, 4, 6]. That is why we shall, furthermore, need case by case arguments in the application of the general theorem to concrete problems.

Some incomplete related results were appeared in [1] as a preliminary report, and the material in this paper is, in part, contained in the thesis of the first author.

2. MAIN THEOREM

For any Hilbert (possibly finite dimensional) space H, let T denote a bounded linear operator of H_K into a Hilbert space H. For the adjoint operator T^* of T, we form the kernel form

$$k(p,q) = (T^*TK(\cdot,q), T^*TK(\cdot,p))_{H_K}, p, q \in E.$$

Note that $k(p,q)$ is a positive matrix on E in the sense of Aronszajn - Moore. Hence, there exists a uniquely determined Hilbert space H_k admitting the reproducing kernel $k(p,q)$. See [2] or [11] for the general properties of reproducing kernels.

For the best approximation stated in the Introduction, we obtain a general and abstract

THEOREM 2.1. *For a member* $\mathbf{g} \in H$, *there exists a function* $\tilde{f}$ *in* H_K *such that*

$$\inf_{f \in H_K} \|Tf - \mathbf{g}\|_H = \|T\tilde{f} - \mathbf{g}\|_H \tag{2.1}$$

if and only if $T^*\mathbf{g}$ *belongs to* H_k. *Furthermore, if the existence of the functions* $\tilde{f}$ *satisfying (2.1) is ensured, then there exists a unique extremal function* $\check{f}$ *with the minimum norm in* H_K, *and the function* $\check{f}$ *is expressible in the form*

$$\check{f}(p) = (T^*\mathbf{g}, T^*TK(\cdot,p))_{H_k}, p \in E. \tag{2.2}$$

Proof. For any $f \in H_K, T^*Tf$ is expressible in the form, by using the reproducing property of $K(\cdot,p)$ in H_K,

$$[T^*Tf](p) = (T^*Tf, K(\cdot,p))_{H_K} = (f, T^*TK(\cdot,p))_{H_K}.$$

Hence, by Theorem 3.2 in [9] (Theorem 3.2 in [11]) the range of T^*T coincides with the Hilbert space H_k. Let P be the orthogonal projection of H_K onto $H_K \ominus N(T^*T)$, $N(T^*T) = \{f \in H_K; T^*Tf = 0\}$. Then, we have, by the same theorem

$$\|T^*Tf\|_{H_k} = \|Pf\|_{H_K}. \tag{2.3}$$

We assume that the extremal functions $\tilde{f}$ satisfying (2.1) exist. Then, we have,

$$\|T\tilde{f} - \mathbf{g}\|_H \leqq \|\mathbf{h} - \mathbf{g}\|_H$$

for all $\mathbf{h}$ in $\overline{R(T)}$ (H - closure of the range of T). Hence, $\mathbf{g} = T\tilde{f} + \mathbf{d}$ for some $\mathbf{d} \in H \ominus \overline{R(T)}$. Since $N(T^*) = H \ominus \overline{R(T)}, T^*T\tilde{f} = T^*\mathbf{g}$, and so, $T^*\mathbf{g}$ is a member of H_k.

To show the converse, let f_1 be any member of H_K with $T^*Tf_1 = T^*\mathbf{g}$. In addition, we choose a member $\mathbf{h}_1$ in $\overline{R(T)}$ such that

$$\|\mathbf{h}_1 - \mathbf{g}\|_H \leqq \|\mathbf{h} - \mathbf{g}\|_H \tag{2.4}$$

for all $\mathbf{h} \in \overline{R(T)}$. Then,

$$T^*Tf_1 = T^*\mathbf{h}_1$$

and

$$Tf_1 = \mathbf{h}_1$$

because T^* is one-to-one on $\overline{R(T)}$. Thus, we have, from (2.4)

$$\|Tf_1 - \mathbf{g}\|_H = \inf_{f \in H_K} \|Tf - \mathbf{g}\|_H.$$

Setting $\check{f} = Pf_1$, we also see that $\check{f}$ is a unique member in H_K such that

$$\|T\check{f} - \mathbf{g}\|_H = \inf_{f \in H_K} \|Tf - \mathbf{g}\|_H,$$

and $\check{f}$ has the minimum norm in H_K because the set of functions f_1 satisfying (2.1) is exactly $\check{f} + N(T^*T)$.

Next, we shall derive the expression (2.2). Since T^*T is an isometry of $H_K \ominus N(T^*T)$ into H_k, its adjoint S is the inverse of T^*T. Hence, we have

$$\begin{aligned}\check{f}(p) &= [ST^*\mathbf{g}](p) \\ &= (ST^*\mathbf{g}, K(\cdot, p))_{H_K} \\ &= (T^*\mathbf{g}, T^*TK(\cdot, p))_{H_k}.\end{aligned}$$

In Theorem 2.1, note that

$$T^*\mathbf{g}(p) = (T^*\mathbf{g}, K(\cdot, p))_{H_K} = (\mathbf{g}, TK(\cdot, p))_H;$$

that is, $T^*\mathbf{g}$ is expressible in terms of $\mathbf{g}$, T and H, concretely. Therefore, when the realization of the norm in the reproducing kernel Hilbert space H_k can be given effectively by the positive matrix $k(p, q)$ which is computable in terms of $K(p, q)$ and H, then we can apply our general theorem to concrete problems. When H is a finite dimension, we can realize the norm in H_k, directly and so, we can apply our theorem in any reproducing kernel Hilbert space H_K. See [2] and [8] for the norm in a finite dimensional reproducing kernel space. For an infinite dimensional case of H, the realization of the norm in H_k is, in general, complicated and we shall need case by case arguments in this realization. In the following part, we shall give a typical and simple result deduced from our theorem.

3. APPROXIMATION BY SOLUTIONS OF HEAT EQUATION

Let $u_F(x,t), 0 < t$, denote the solution of the Cauchy problem :

$$u_{xx}(x,t) = u_t(x,t) \quad \text{on} \quad \mathbb{R} \times \{t > 0\} \tag{3.1}$$

and

$$u(x,0) = F(x) \quad \text{on} \quad \mathbb{R}, \tag{3.2}$$

where $F \in L_2(\mathbb{R}, dx)$. Then we have $u_F(x,t) \in L_2(\mathbb{R}, dx)$ (see Theorem 3.2, below).
Given $h \in L_2(\mathbb{R}, dx)$, with fixed t, we want to minimize expression :

$$\int_{\mathbb{R}} |u_F(x,t) - h(x)|^2 dx,$$

where F runs over $L_2(\mathbb{R}, dx)$. Then, we can obtain the following results.

First, we determine a necessary and sufficient condition on h for which there exists $F \in L_2(\mathbb{R}, dx)$ with

$$\int_{\mathbb{R}} |u_F(x,t) - h(x)|^2 dx = 0.$$

Secondly, we will construct, for any given h in $L_2(\mathbb{R}, dx)$, the minimizing sequence $\{u_{F_n}\}$ for which

$$\lim_{n\to\infty} \int_{\mathbb{R}} |u_{F_n}(x,t) - h(x)|^2 dx = 0.$$

We shall first characterize the Hilbert space formed by the solutions $u_F(x,t)$ of (3.1) and (3.2) using the theory of reproducing kernels ([9] and [11]).

THEOREM 3.1 ([10]). *For any fixed $t > 0$, the solutions of (3.1) - (3.2) are extensible analytically onto $\mathbb{C}$ in the form $u_F(z,t)$ and form the reproducing kernel Hilbert space $H_{K(t)}$ with the norm*

$$\|u_F(\cdot,t)\|^2_{H_{K(t)}} = \frac{1}{\sqrt{2\pi t}} \iint_{\mathbb{C}} |u_F(z,t)|^2 \exp\{-\frac{y^2}{2t}\} dx dy,$$

where the reproducing kernel $K(z,\overline{u};t)$ in $H_{K(t)}$ is given by

$$K(z,\overline{u};t) = \frac{1}{2\sqrt{2\pi t}} \exp\{-\frac{z^2}{8t}\} \exp\{-\frac{\overline{u}^2}{8t}\} \exp\{\frac{z\overline{u}}{4t}\}.$$

Furthermore, the initial functions F are given by

$$F(\xi) = s - \lim_{N\to\infty} \frac{1}{\sqrt{2\pi t}} \iint_{|z|<N} u_F(z,t) \overline{k(z-\xi,t)} \exp\{-\frac{y^2}{2t}\} dx dy,$$

where, the limit is taken over $L_2(\mathbb{R}, d\xi)$ and

$$k(z,t) = \frac{1}{\sqrt{4\pi t}} \exp\{-\frac{z^2}{4t}\}.$$

We define the linear operator $T : H_{K(t)} \longrightarrow L_2(\mathbb{R}, dx)$ by the restriction Tf of f to $\mathbb{R}$. Then, the following inequality of Fejér-Riesz type implies that T is well-defined and is bounded on the space $H_{K(t)}$.

THEOREM 3.2. *For any $f \in H_{K(t)}$, the following inequality holds:*

$$\int_{\mathbb{R}} |f(x)|^2 dx \leqq \|f\|^2_{H_{K(t)}}.$$

Here, the constant 1 *as the coefficient of $\|f\|^2_{H_{K(t)}}$ is best possible in the inequality.*

The reproducing kernel $k(z, \overline{u}; t)$ determining the range T^*T is given by

$$\begin{aligned} k(z, \overline{u}; t) &= (T^*TK(\cdot, \overline{u}; t), T^*TK(\cdot, \overline{z}; t))_{H_{K(t)}} \\ &= \frac{1}{2\sqrt{6\pi t}} \exp\{-\frac{z^2}{24t}\} \exp\{-\frac{\overline{u}^2}{24t}\} \exp\{\frac{z\overline{u}}{12t}\} \end{aligned}$$

and, the Hilbert space $H_{k(t)}$ admitting the reproducing kernel $k(z, \overline{u}; t)$ is composed of all entire functions $f(z)$ with finite norms

$$\|f\|^2_{H_{k(t)}} = \frac{1}{\sqrt{6\pi t}} \iint_{\mathbb{C}} |f(z)|^2 \exp\{\frac{-y^2}{6t}\} dxdy.$$

Then, by using Theorem 2.1 we obtain

THEOREM 3.3. *For $h \in L_2(\mathbb{R}, dx)$, there exists a member $F \in L_2(\mathbb{R}, dx)$ such that*

$$u_F(x, t) = h(x) \quad on \quad \mathbb{R}$$

if and only if

$$\iint_{\mathbb{C}} \left| \int_{\mathbb{R}} h(\xi) \exp\{-\frac{\xi^2}{8t} + \frac{z\xi}{4t}\} d\xi \right|^2 \cdot \exp\{\frac{-3x^2 + y^2}{12t}\} dxdy < \infty. \tag{3.3}$$

For any $h \in L_2(\mathbb{R}, dx)$, we shall construct a sequence $\{u_n(z)\}$ such that

$$u_n(z) \in H_{K(t)}$$

and

$$\lim_{n \to \infty} \int_{\mathbb{R}} |u_n(x) - h(x)|^2 dx = 0.$$

First, note that the images

$$f(z) = (T^*g)(z) = (g(\cdot), TK(\cdot, \overline{z}; t))_{L_2(\mathbb{R}, dx)}$$

of the members g in $L_2(\mathbb{R}, dx)$ are characterized as the reproducing kernel Hilbert space $H_{\mathbf{K}(t)}$ whose reproducing kernel is

$$\mathbb{K}(z, \overline{u}; t) = (TK(\cdot, \overline{u}; t), TK(\cdot, \overline{z}; t))_{L_2(\mathbb{R}, dx)}.$$

From the concrete expression of $\mathbb{K}(z, \overline{u}; t)$

$$[T^*TK(\cdot, \overline{u}; t)](z) = \frac{1}{4\sqrt{\pi t}} \exp\{-\frac{z^2}{16t}\} \exp\{-\frac{\overline{u}^2}{16t}\} \exp\{\frac{z\overline{u}}{8t}\},$$

we see that the family of functions

$$\varphi_n(z) = \frac{1}{2}[n!(8t)^n \sqrt{\pi t}]^{-\frac{1}{2}} \exp\{-\frac{z^2}{16}\} z^n,$$
$$(n = 0, 1, 2, \cdots)$$

is a complete orthonormal system in $H_{\mathbf{K}(t)}$. Note that

$$\varphi_n \in H_{k(t)} \qquad (n = 0, 1, 2, \cdots).$$

We set

$$a_n = (T^*h, \varphi_n)_{H_{\mathbf{K}(t)}}$$

and

$$\hat{f}_N(z) = \sum_{n=0}^{N} a_n \varphi_n(z).$$

Then, $\hat{f}_N \in H_{k(t)} \subset H_{\mathbf{K}(t)}$ and the sequence $\{\hat{f}_N\}$ converges to T^*h in $H_{\mathbf{K}(t)}$ as N tends to infinity. As in the proof of Theorem 2.1, we construct the sequence u_N^* satisfying

$$T^*Tu_N^* = \hat{f}_N.$$

Since the adjoint operator T^* is an isometry of $L_2(\mathbb{R}, dx)$ onto $H_{\mathbf{K}(t)}$, we obtain

THEOREM 3.4. *For any $h \in L_2(\mathbb{R}, dx)$, let the sequence $\{u_N^*\}_{N=0}^{\infty}$ be constructed by the above method. Then,*

$$\lim_{N\to\infty} \int_{\mathbb{R}} |[Tu_N^*](x) - h(x)|^2 dx = 0.$$

ACKNOWLEDGMENT

The travel and living expenses of the second author for attending to the Second International Colloquium on Numerical Analysis, Plovdiv, August 13-17, 1993 were supported by the Japanese Ministry of Education, Science and Culture.

REFERENCES

1. T. Ando and S. Saitoh, Suri Kaiseki Kenkyu Jo, Koukyu Roku, 743, 164-187 (1991).
2. N. Aronszajn, Trans. Amer. Math. Soc. 68, 337-404(1950).
3. D. -W. Byun, Mathematica Japonica 38, 477-481(1993).
4. D. -W. Byun, Proc. Amer. Math. Soc. 118, 437-445 (1993).
5. D. -W. Byun, Integral Transforms and Approximation of Functions , Thesis, Gunma University (1993).
6. D. -W. Byun and S. Saitoh, Analytic extensions of functions on the real line to entire functions, preprint.
7. D. -W. Byun and S. Saitoh, Approximation by the solutions of the heat equation, J. Approximation Theory (in press).
8. S. Saitoh, Linear Algebra Appl. 48, 119-130 (1982)
9. S. Saitoh, Proc. Amer. Math. Soc. 89, 74-78 (1983).
10. S. Saitoh, Applicable Analysis 16, 1-6 (1983).
11. S. Saitoh, Theory of Reproducing Kernels and its Applications, Pitman Res. Notes in Math. Series 189, Longman Science & Technical, England, (1988).

2nd Int. Coll. on Numerical Analysis, pp. 63-72
D. Bainov and V. Covachev (Eds)

Chebyshev Spectral Solution of the Burgers Equation with High Wavenumbers

H. DANG-VU
Institut de Mathématiques pures et appliquées
Université Pierre et Marie Curie, 75252-Paris. France

C. DELCARTE
LIMSI-Université de Paris-Sud, 91405-Orsay. France

August 16, 1993

Abstract

A new procedure is introduced to solve efficiently the full matrix equation resulting from spectral approximations of the Burgers equation. The method considered here is an extension of that developed by us in [3]. It is shown that the accuracy of the Chebyshev Spectral solution of the Burgers equation decreases as the wavenumber increases. A correlation equation for critical number of collocation points as a function of viscosity is proposed.

Keywords : Chebyshev Spectral Method, Burgers Equation.
AMS(MOS) Subject Classifications : 65N35, 65M10

1 Introduction

The problem to be studied here is the Burgers Equation :

$$\begin{aligned} \frac{\partial u}{\partial t}(x,t) &= -u(x,t)\frac{\partial u}{\partial x}(x,t) + \nu\frac{\partial^2 u}{\partial x^2}(x,t) \\ -1 \leq x &\leq 1, \quad t \geq 0, \quad \nu > 0 \end{aligned} \tag{1.1}$$

with homogeneous Dirichlet boundary conditions :

$$u(-1,t) = 0, \quad u(1,t) = 0 \tag{1.2}$$

and initial condition :

$$u(x,0) = \sin k\pi x \qquad k = 1,2,3,\ldots \tag{1.3}$$

System (1.1)-(1.3) with $\nu = 1/100\pi$ and $k = 1$ has been solved by Basdevant et al [1] using a variety of methods. The conclusion drawn by them is that spectral methods do yield better accuracy than other methods (Basdevant et al actually prove their results for $u(x,0) = -\sin \pi x$). However it is well known that the approximation of the *higher* eigenfunctions of the second order derivative by polynomial interpolation is not accurate [5]. As a consequence, we expect that there may be difficulty in obtaining an accurate approximation to the solution of the system (1.1)-(1.3) at large k by spectral method.

The accuracy of the spectral solution of (1.1)-(1.3) will be evaluated by comparison with the exact solution obtained by means of the Cole-Hopf transformation [4] :

$$u(x,t) = 4k\pi\nu \ \sum_{n=1}^{\infty} \ \exp(-n^2k^2\pi^2\nu t) n I_n(\frac{1}{2k\pi\nu}) \sin nk\pi x \Big/ \sum_{n=0}^{\infty} \ \frac{2}{c_n}\exp(-n^2k^2\pi^2\nu t) I_n(\frac{1}{2k\pi\nu}) \cos nk\pi x$$

where $c_0 = 2$, $c_n = 1$ if $n > 0$ and $I_n(z)$ is the modified Bessel function of the first kind.

The interest of this investigation goes far beyond the Burgers equation. Indeed, similar considerations apply to more complicated problems, such as the Navier-Stokes equations.

2 A Chebyshev Spectral Method for the Burgers Equation

Equation (1.1) can be efficiently computed in the spectral space by means of the transform :

$$u_N(x,t) = \sum_{k=0}^{N} a_k(t) T_k(x) \tag{2.1}$$

where $T_k(x)$ denotes the Chebyshev polynomial of degree k defined by

$$T_k(\cos\theta) = \cos(k\theta) \text{ with } x = \cos\theta \tag{2.2}$$

The set of Chebyshev polynomials $T_k(x)$ is an orthogonal system with respect to the inner product defined by :

$$< T_i, T_j > = \int_{-1}^{1} T_i(x) T_j(x) \frac{dx}{\sqrt{1-x^2}} \tag{2.3}$$

namely :

$$
\begin{aligned}
< T_i, T_j > &= 0, \text{ if } i \neq j \\
< T_i, T_i > &= c_i \frac{\pi}{2}, \qquad i = 0, 1, 2, \ldots
\end{aligned} \tag{2.4}
$$

where $c_0 = 2$ and $c_i = 1$ for $i > 0$.

Each term in the differential equation (1.1) is expanded as a truncated series of Chebyshev polynomials as :

$$\frac{\partial u_N(x,t)}{\partial x} = \sum_{k=0}^{N} a_k^{(1)}(t) T_k(x) \tag{2.5}$$

$$\frac{\partial^2 u_N(x,t)}{\partial x^2} = \sum_{k=0}^{N} a_k^{(2)}(t) T_k(x) \tag{2.6}$$

$$u_N(x,t) \frac{\partial u_N(x,t)}{\partial x} = \sum_{k=0}^{N} b_k(t) T_k(x) \tag{2.7}$$

where [2] :

$$a_k^{(1)}(t) = \frac{2}{c_k} \sum_{\substack{p=k+1 \\ p+k \text{ odd}}}^{N} a_p(t), \qquad a_k^{(2)}(t) = \frac{1}{c_k} \sum_{\substack{p=k+2 \\ p+k \text{ even}}}^{N} p(p^2 - k^2) a_p(t) \tag{2.8}$$

and

$$
\begin{aligned}
b_k(t) &= \tfrac{2}{\pi c_k} \int_{-1}^{1} u_N(x,t) \tfrac{\partial u_N(x,t)}{\partial x} T_k(x) \tfrac{dx}{\sqrt{1-x^2}} \\
&= \tfrac{2}{\pi c_k} \sum_{i,j=0}^{N} a_i(t) a_j^{(1)}(t) \int_{-1}^{1} T_i(x) T_j(x) T_k(x) \tfrac{dx}{\sqrt{1-x^2}}
\end{aligned}
$$

Due to the relation (2.2) we have

$$
\begin{aligned}
\int_{-1}^{1} T_i(x) T_j(x) T_k(x) \tfrac{dx}{\sqrt{1-x^2}} &= \int_0^{\pi} \cos i\theta \cos j\theta \cos k\theta \, d\theta \\
&= \tfrac{\pi}{4} (\delta_{i+j,k} + \delta_{i+j,-k} + \delta_{i-j,k} + \delta_{i-j,-k})
\end{aligned}
$$

It follows that

$$b_k(t) = \frac{1}{2c_k} \sum_{i,j=0}^{N} a_i(t) a_j^{(1)}(t) (\delta_{i+j,k} + \delta_{i+j,-k} + \delta_{i-j,k} + \delta_{i-j,-k}) \tag{2.9}$$

Equation (1.1) reduces to :

$$\frac{da_k(t)}{dt} = -b_k(t) + \nu a_k^{(2)}(t) \qquad k = 0, 1, \ldots N \tag{2.10}$$

where $a_k^{(2)}$ and $b_k(t)$ are given by (2.8) and (2.9) respectively.

The boundary conditions (1.2) become, through the use of (2.1)

$$\sum_{\substack{p=0 \\ p \text{ even}}}^{N} a_p(t) = 0, \qquad \sum_{\substack{p=1 \\ p \text{ odd}}}^{N} a_p(t) = 0 \tag{2.11}$$

We look for a tau solution of (2.10)-(2.11). The idea of the tau method (see e. g. [2]) is to apply (2.10) for $n = 0, 1, 2, \ldots N-2$ only. The remaining equations are obtained by imposing the boundary conditions (2.11).

3 Time-discretization

The numerical solution of (2.10) in time, for $k = 0, 1, \ldots, N-2$ can be done using any numerically stable time integration method for ordinary differential equations, such as the Adams-Bashforth Crank-Nicolson (ABCN) method.

The ABCN approximation to (2.10) reads as follows :

$$\begin{aligned} \frac{[a_k]^{n+1} - [a_k]^n}{\Delta t} &= \frac{\nu}{2}([a_k^{(2)}]^{n+1} + [a_k^{(2)}]^n) - \frac{1}{2}(3[b_k]^n - [b_k]^{n-1}) \\ k &= 0, 1, 2, \ldots N-2 \end{aligned} \tag{3.1}$$

where the superscripts denote the time-level at which the term is evaluated. This scheme combines an Adams-Bashforth method for the non-linear term and a Crank-Nicolson method for the linear term.

Therefore , at each time-step, we have to solve a Helmholtz equation :

$$\frac{1}{c_k} \sum_{\substack{p=k+2 \\ p+k \text{ even}}}^{N} p(p^2 - k^2)[a_p]^{n+1} - \kappa [a_k]^{n+1} = g_k \quad k = 0, 1, \ldots N-2 \tag{3.2}$$

with $\kappa = 2/\nu\Delta t$ and

$$g_k = -([a_k^{(2)}]^n + \kappa[a_k]^n) + \frac{1}{\nu}(3[b_k]^n - [b_k]^{n-1})$$

is a known vector of previously computed values at time levels $n\Delta t$ and $(n-1)\Delta t$.

For convenience we suppose that N is even and $N \geq 6$.

a) Case $\kappa = 0$:

For $\kappa = 0$ the system (3.2) gives the spectral equations of the Poisson equation

$$\frac{1}{c_k} \sum_{\substack{p=k+2 \\ p+k \text{ even}}}^{N} p(p^2 - k^2)a_p = g_k, \quad 0 \leq k \leq N-2 \tag{3.3}$$

and we have shown in [3] that in this case the exact solution to (3.3), with the boundary conditions (2.11), is :

$$a_i = r_i g_{i-2} + s_i g_i + v_i g_{i+2} \qquad 2 \leq i \leq N \tag{3.4}$$

where

$$r_i = \frac{c_{i-2}}{4i(i-1)} \quad s_i = -\frac{e_{i+2}}{2(i^2-1)} \quad v_i = \frac{e_{i+4}}{4i(i+1)} \quad 2 \leq i \leq N \tag{3.5}$$

Here $e_i = 1$ for $i \leq N$, $e_i = 0$ for $i > N$ and a_0, a_1 are given by (2.11) :

$$\begin{aligned} a_0 &= -(a_2 + a_4 + \cdots + a_N) \\ a_1 &= -(a_3 + a_5 + \cdots + a_{N-1}) \end{aligned} \tag{3.6}$$

Let P be the upper triangular $(N+1)\times(N+1)$ matrix

$$\begin{pmatrix} 1 & 0 & 1 & 0 & 1 & \dots & 1 \\ & 1 & 0 & 1 & 0 & \dots & 0 \\ & & 2^3/2 & 0 & 4^3/2 & \dots & N^3/2 \\ & & & 3(3^2-1^2) & 0 & \dots & 0 \\ & & & & & & \vdots \\ & & & & & \dots & 4N(N-1) \end{pmatrix}$$

and a, g be column vectors of length $N+1$ defined by :

$$\begin{aligned} a^T &= (a_0, a_1, a_2, \dots, a_N) \\ g^T &= (0, 0, g_0, g_1, \dots, g_{N-2}) \end{aligned}$$

then (3.3) in combination with (2.11) can be written as :

$$Pa = g \tag{3.7}$$

and (3.4)-(3.6) is the solution to (3.7).

b) Case $\kappa \neq 0$ *:*

If we put

$$\bar{g}^T = (0, 0, g_0 + \kappa a_0, g_1 + \kappa a_1, \dots, g_{N-2} + \kappa a_{N-2}) \tag{3.8}$$

then (3.2) becomes

$$Pa = \bar{g} \tag{3.9}$$

The equation (3.9) has the same structure as the equation (3.7) so that we can write (3.9) as

$$a_i = r_i \bar{g}_{i-2} + s_i \bar{g}_i + v_i \bar{g}_{i+2} \qquad 2 \leq i \leq N \tag{3.10}$$

$$a_N = -(a_0 + a_2 + \dots + a_{N-2}) \tag{3.11}$$

$$a_{N-1} = -(a_1 + a_3 + \dots + a_{N-3}) \tag{3.12}$$

From (3.8) and (3.9) we obtain

$$-\kappa r_i a_{i-2} + (1 - \kappa s_i) a_i - \kappa v_i a_{i+2} = h_i \quad 2 \leq i \leq N \tag{3.13}$$

where

$$h_i = r_i g_{i-2} + s_i g_i + v_i g_{i+2} \qquad 2 \leq i \leq N$$

Note that the even and odd coefficients are uncoupled in (3.11)-(3.13). Let $w \in \{0,1\}$ then (3.11)-(3.13) can be written as

$$\begin{pmatrix} 1 & 1 & 1 & \dots & 1 & 1+\kappa r_{N-w} \\ -\kappa r_{2+w} & 1-\kappa s_{2+w} & -\kappa v_{2+w} & & & \\ & \dots & \dots & \dots & & \\ & & -\kappa r_{2i+w} & 1-\kappa s_{2i+w} & -\kappa v_{2i+w} & \\ & & \dots & \dots & \dots & \\ & & & & -\kappa r_{N-2-w} & 1-\kappa s_{N-2-w} \end{pmatrix} \times$$

$$\begin{pmatrix} a_w \\ a_{2+w} \\ \vdots \\ a_{2i+w} \\ \vdots \\ a_{N-2-w} \end{pmatrix} = \begin{pmatrix} -h_{N-w} \\ h_{2+w} \\ \vdots \\ h_{2i+w} \\ \vdots \\ h_{N-2-w} \end{pmatrix} \quad \text{with w=0,1} \qquad (3.14)$$

The linear system (3.14) may be solved by the Thomas-Cholesky method or by the Gauss elimination procedure. In the Appendix a computation of the solution to (3.14) based on the Thomas-Cholesky method is given. Finally A_N and A_{N-1} are given by (3.11) and (3.12) respectively.

4 Numerical Results

All the calculations were realized using the Cray 2 computer. In Table I we list critical values N_c of N corresponding to $k = 1$. If $N < N_c$ there may be rapid growth of errors and numerical solution of (3.8)-(3.9) may appear unstable and divergent. N_c seems to grow at most like

$$N_c \simeq \frac{c}{\sqrt{\nu}} \text{ with } c \simeq 1.3 \qquad (4.1)$$

TABLE I
Critical value N_c ($k = 1$) as a function of ν
The number $c/\sqrt{\nu}$ is also listed.

ν	$1/100\pi$	$1/200\pi$	$1/300\pi$	$1/400\pi$
N_c	23	33	40	46
$c\nu^{-\frac{1}{2}}$	23	33.5	39.9	46

We report in Figures 1 through 4 some numerical experiments for different selected values of k and N. The time step $\Delta t = 10^{-4}/\pi$ was chosen so small that time differencing errors are negligible. For smaller Δt values, the errors remained constant. All runs were made from $t = 0$ to $t = 3/10\pi$ (3000 time-steps). It should be noted that at $t = 3/10\pi$ the sawtooth profile is observed for all $k \geq 10$. In Figures 1 and 2 we compare the spectral solutions for $N = 60$, $\nu = 1/160\pi$, $k = 2, 3$ with the exact solutions. Figures 1a and 2a represent the exact solutions. The corresponding polynomial interpolants $u_N(x,t)$ are represented in Figures 1b and 2b respectively (The circles denote the collocation points x_i). The plot in Figure 1b demonstrates the convergence behavior of the spectral method for $k = 2$. The breakdown in convergence at wavenumbers above 2 is clearly observed in Figure 2b. In Figure 3 we show the plot of $u_N(x,t)$ at $k = 10$ for various values of N. The computations were carried out for $\nu = 1/60\pi$. Observe that the approximate solutions converge to the exact solution as N increases. Approximation is accurate for roughly $N = 180$.

Now we define $N_c(k)$, with $N_c(1) = N_c$, as the critical number for the wavenumber k, in a practical manner as follows : for a fixed k, if $N < N_c(k)$ then the numerical results exhibit oscillations that grow very rapidly, and after a few time-steps their amplitude is infinite so that an "overflow" is registered by the computer. Figure 4 shows the critical number $N_c(k)$ as a function of k for each of four viscosities considered. This result leads to an heuristic rule for $N_c(k)$:

$$N_c(k) = kN_c \tag{4.2}$$

where N_c is given in Table I.

Figure 1 Comparison of exact and Chebyshev Spectral solutions at 3000 time-steps, $\Delta t = 10^{-4}/\pi$ for $\nu = 1/160\pi$, $k = 2$, $N = 60$. (a) Exact solution, (b) Chebyshev Spectral solution

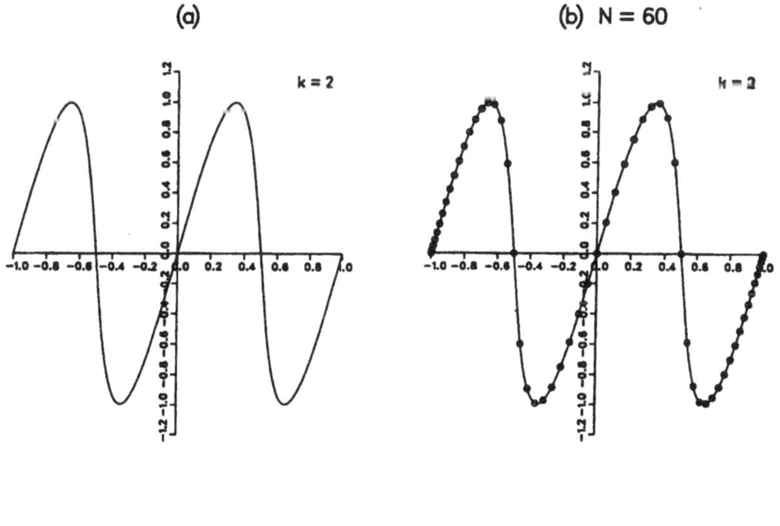

Figure 2 Comparison of exact and Chebyshev Spectral solutions at 3000 time-steps, $\Delta t = 10^{-4}/\pi$ for $\nu = 1/160\pi$, $k = 3$, $N = 60$. (a) Exact solution, (b) Chebyshev Spectral solution

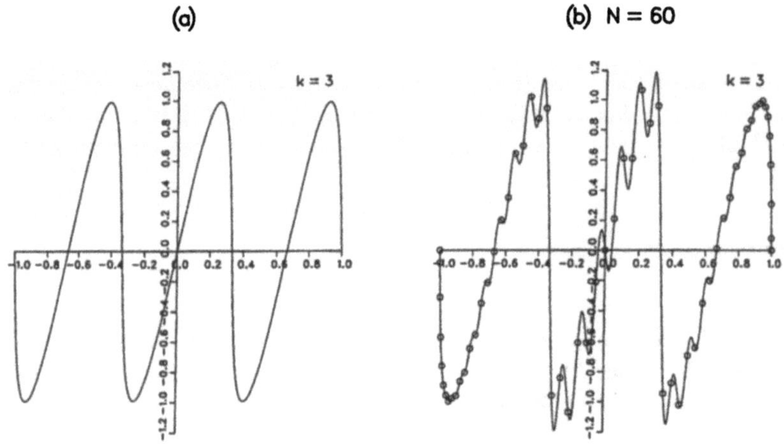

Figure 3 A plot of the Chebyshev Spectral solutions at $k = 10$ for $\nu = 1/60\pi$, $N = 50$, $N = 100$ and $N = 180$. The continuous line represents the exact solution.

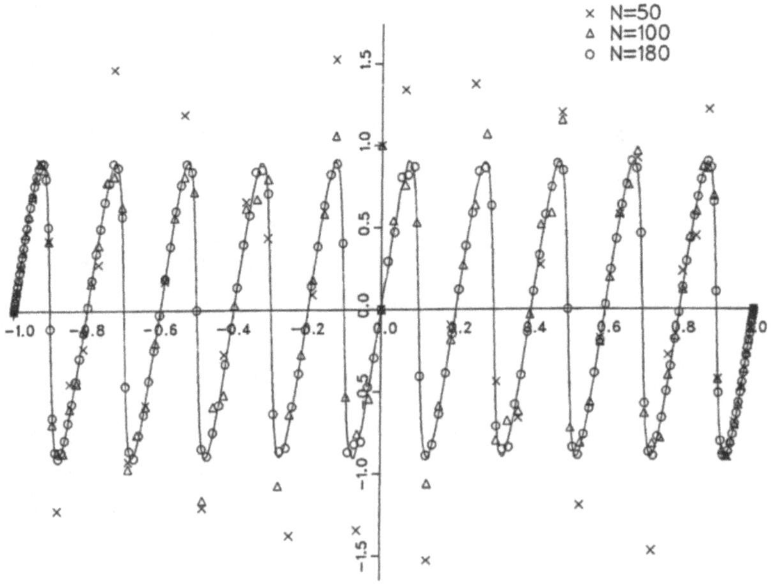

Figure 4 Critical number $N_c(k)$ as a function of k for $\nu = 1/100\pi$, $\nu = 1/200\pi$, $\nu = 1/300\pi$ and $\nu = 1/400\pi$.

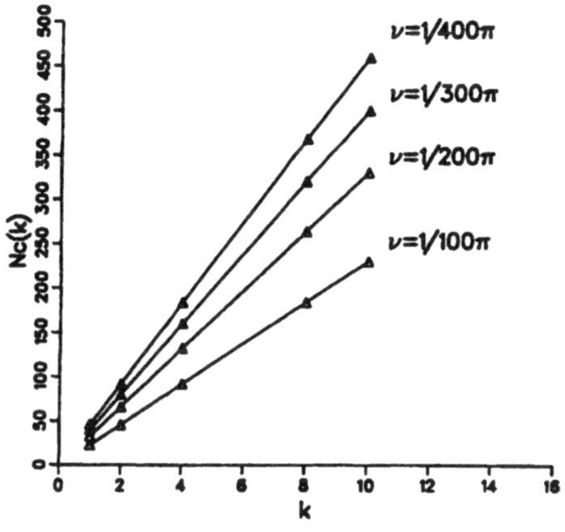

APPENDIX

Computation of the direct solution to (3.14)

The system (3.14) can be written as :

$$a_w + a_{2+w} + \cdots + a_{N-4-w} + (1 + \kappa r_{N-w}) a_{N-2-w} = -h_{N-w} \qquad (A.1)$$

$$-\kappa r_{N-2i-w} a_{N-2i-2-w} + (1 - \kappa s_{N-2i-w}) a_{N-2i-w} \qquad (A.2)$$

$$-\kappa v_{N-2i-w} a_{N-2i+2-w} = h_{N-2i-w} \quad 2 \leq i \leq \frac{1}{2}N - 1$$

$$-\kappa r_{N-2-w} a_{N-4-w} + (1 - \kappa s_{N-2-w}) a_{N-2-w} = h_{N-2-w} \qquad (A.3)$$

We seek recursion relations of the form :

$$a_{N-2i+2-w} = \alpha_{2i-2+w} a_{N-2i-w} + \beta_{2i-2+w}$$

Substituting the above expression into (A.2) we obtain :

$$a_{N-2i-w} = \alpha_{2i+w} a_{N-2i-2-w} + \beta_{2i+w} \qquad (A.4)$$

$$\alpha_{2i+w} = \frac{k r_{N-2i-w}}{1 - k s_{N-2i-w} - k v_{N-2i-W} \alpha_{2i+w-2}} \qquad (A.5)$$

$$\beta_{2i+w} = \frac{h_{N-2i-w} + k v_{N-2i-w} \beta_{2i+w-2}}{1 - k s_{N-2i-w} - k v_{N-2i-w} \alpha_{2i+w-2}} \qquad (A.6)$$

$$1 \leq i \leq \tfrac{1}{2}N - 1$$

(A.3) shows that :

$$\alpha_w = \beta_w = 0 \qquad (A.7)$$

Using the recursion relations (A.4) we can write (A.1) as

$$\begin{aligned}\gamma_{N-2+w}a_w &+ \gamma_{N-4+w}\beta_{N-2+w} + \gamma_{N-6+w}\beta_{N-4+w} + \cdots \\ &+ \gamma_{2+w}\beta_{4+w} + \gamma_w\beta_{2+w} = -h_{N-w}\end{aligned} \qquad (A.8)$$

Here γ_{2i+w} are defined by the recursion relations

$$\begin{aligned}\gamma_w &= 1 + kr_{N-w} \\ \gamma_{2i+w} &= 1 + \gamma_{2i-2+w}\alpha_{2i+w}, \quad 1 \le i \le \tfrac{1}{2}N - 1\end{aligned} \qquad (A.9)$$

$a_{2+w}, a_{4+w}, \ldots a_{N-2-w}$ are then given by (A.4).

Thus the algorithm for the solution to (3.14) is :

1) Compute the constants α_{2i+w} and β_{2i+w}, $1 \le i \le \frac{1}{2}N - 1$ using (A.5) and (A.6) respectively.

2) Compute the constants γ_{2i+w}, $1 \le i \le \frac{1}{2}N - 2$ using (A.9)

3) Compute a_w using (A.8).

4) Compute $a_{2+w}, a_{4+w}, \ldots, a_{N-2-w}$ using the recurrence relations (A.4).

References

[1] C. Basdevant, M. Deville, P. Haldenwang, J. M. Lacroix, J. Ouazzani, J. Peyret, R. Orlandi and P. Patera, *Comput. Fluids* **14**, 23-41 (1979).

[2] C. Canuto, M. Y. Hussaini, A. Quarteroni and T. Zang, *Spectral Methods in Fluid Dynamics*, Springer, Berlin, (1977).

[3] H. Dang-Vu and C. Delcarte, *J. Comput. Phys* **104**, 211-220 (1993).

[4] P. L. Sachdev, *Nonlinear Diffusive Waves*, Cambridge University Press, Cambridge, (1987).

[5] J. A. C. Weideman and L. N. Trefethen, *SIAM J. Numer. Anal.* **25**, 1279-1298 (1988).

2nd Int. Coll. on Numerical Analysis, pp. 73-82
D. Bainov and V. Covachev (Eds)

PADÉ APPROXIMATION AND NUMERICAL INVERSION OF THE LAPLACE TRANSFORM

Pablo González-Vera and Ramón Orive

Departamento de Análisis Matemático, Universidad de La Laguna
Canary Islands, Spain

Abstract

In this paper we aim to study different approaches to get numerical inversion of the Laplace Transform by using Padé-type Approximation in one or two points. Results concerning with convergence are discussed.

Keywords. Padé-type Approximants, Laplace Transform, asymptotic expansions, uniform convergence.

1. INTRODUCTION

As is well known, a usual problem in the applications of the Laplace transform is to determine the original function f(t), assuming that the transform function, given by $L\{f(t)\} = \bar{f}(s) = \int_0^\infty e^{-st}f(t)dt$ is known. In this sense, the immediate inversion of rational functions suggests the use of Padé Approximants (PAs) (see e.g. [1]) to solve this problem.

Indeed, if $\bar{f}_m(s) = Q_{m-1}(s)/P_m(s)$ is a [m-1/m] PA to $\bar{f}$ (we take [m-1/m] PAs because of $\lim_{Re(s)\to\infty} \bar{f}(s) = 0$), by partial fractions decomposition we can write $\bar{f}_m(s) = \sum_{i=1}^{m} \sum_{k=0}^{\beta_i - 1} \gamma_{ik} (x-x_i)^{-(k+1)}$, where x_i are distincts and $\sum_{i=1}^{m} \beta_i = n$. Therefore, $L^{-1}\{\bar{f}_m(s)\} = f_m(t) = \sum_{i=1}^{m} \sum_{k=0}^{\beta_i - 1} \gamma_{ik} t^k e^{x_i t}$ (as usual L^{-1} denotes the inverse Laplace Transform). So, the use of PA's to obtain the inversion of Laplace Transform consists of approximating the original function by functions f_m as given above.

A natural drawback of this method is that the poles of PA's may be located on the half-plane of convergence. Furthermore, such these poles could be not easily computable. In order to solve this problem, the use of

Padé Approximants with fixed poles (the so-called Padé-type Approximants or abreviately 1PTAs, analyzed by Brezinski in [2]) seems to be convenient.

On the other hand, since $\lim_{Re(s)\to\infty} \bar{f}(s) = 0$, it is usual that $\bar{f}$ admits an expansion near infinity of the type $\bar{f}(s) \sim \sum_{j=1}^{\infty} d_j s^{-j}$ $(s\to\infty)$. This fact suggests the possibility of making use of Two-point Padé-type Approximants (2PTAs, see e.g. [3]). Namely, given two (possibly formal) power series

$$L_0 \equiv \sum_{j=0}^{\infty} c_j z^j, \quad L_\infty \equiv \sum_{j=1}^{\infty} d_j z^{-j}$$

two nonnegative integers k,m $(0\le k\le m)$ and a polynomial P_m $(P_m(0)\neq 0)$ of degree m, there exists a unique polynomial Q_{m-1} of degree not greater than $(m-1)$, such that the rational function (Q_{m-1}/P_m) satisfies

$$L_0(z) - (Q_{m-1}(z)/P_m(z)) = O(z^k) \quad (z\to 0) \tag{1.1}$$

$$L_\infty(z) - (Q_{m-1}(z)/P_m(z)) = O\left((z^{-1})^{(m-k+1)}\right) \quad (z\to\infty) \tag{1.2}$$

Then, the rational function Q_{m-1}/P_m is said to be a (k/m) two-point Padé-type Approximant to the pair (L_0,L_∞) and denoted by: $\dfrac{Q_{m-1}(z)}{P_m(z)} = (k/m)_{(L_0,L_\infty)}(z)$

In the sequel we shall deal with sequences of such approximants to solve the numerical inversion of the Laplace Transform.

2. ABOUT THE EXISTENCE OF A BEST APPROXIMATION

Let $\bar{f}$ be an analytical function in the half plane $Re(s) > \gamma$. By means of a simple translation, we can suppose that $\gamma < 0$ and therefore $\bar{f}$ is analytic at the origin. So, we can write $\bar{f}(s) = \sum_{j=0}^{\infty} c_j s^j \equiv \bar{f}_0$. On the other hand, assume that $\bar{f}$ admits an asymptotic expansion of the form $\bar{f}(s) \sim \sum_{j=1}^{\infty} d_j s^{-j} \equiv f_\infty$. We consider (k/m)-2PTA's to $\bar{f}$ of the form:

$$\bar{f}_m(s) = (\bar{Q}_{m-1}/\bar{P}_m)(s) = \sum_{l=1}^{n} \sum_{k=0}^{\beta_l -1} \gamma_{lk} (s-s_l)^{-(k+1)} \tag{2.1}$$

where $\sum_{l=1}^{n} \beta_l = m$ and $Re(s_l) < \gamma$. Obviously, we have:

$$L^{-1}\{\bar{f}_m(s)\} = \sum_{l=1}^{n} \sum_{k=0}^{\beta_l -1} \gamma_{lk}\, t^k e^{s_l t} = f_m(t) \tag{2.2}$$

If we assume that

$$\int_0^\infty e^{-\omega t} \{f(t)\}^2\, dt < \infty \tag{2.3}$$

(hence, $f(t) = O(e^{\omega t/2})$, when $t\to\infty$) and

$$\int_0^\infty e^{-\omega t} \, |f(t)| \, dt < \infty \tag{2.4}$$

then, from the Parseval Identity (see e.g. [4]), we have:

$$\int_0^\infty e^{-\omega t} \{f(t)\}^2 \, dt = \frac{1}{2\pi} \int_{\omega/2 - i\infty}^{\omega/2 + i\infty} |\bar{f}(s)|^2 \, ds \tag{2.5}$$

Now, the following questions are going to be solved:

Optimization: find the poles "$\bar{s}_l$" that minimize the cuadratic mean error:

$$I_m(\omega;s_1,\dots,s_n) = \int_0^\infty e^{-\omega t} \{f(t)-f_m(t)\}^2 \, dt$$

Convergence: which conditions ensure that $\lim_{m\to\infty} I_m(\omega;\bar{s}_1,\dots,\bar{s}_n) = 0$?

Let us now consider the classes of functions:

$$\bar{F}_m = \left\{ \bar{f}_m(s) = (k/m)_{(\bar{f}_0,\bar{f}_\infty)}(s) = \sum_{l=1}^{n} \sum_{k=0}^{\beta_l -1} \gamma_{lk} \, (s-s_l)^{-k-1}, \ \mathrm{Re}(s_l) < \omega/2 \right\}$$

$$F_m = \left\{ f_m(t) = L^{-1}\{\bar{f}_m(s)\} = \sum_{l=1}^{n} \sum_{k=0}^{\beta_l -1} \gamma_{lk} \, t^k e^{s_l t}, \ \mathrm{Re}(s_l) < \omega/2 \right\}$$

It is clear that if $\bar{f}_m \in \bar{F}_m$, then f_m satisfies (2.3) and (2.4) and hence

$$\int_0^\infty e^{-\omega t} \{f_m(t)\}^2 \, dt < \infty \ \text{ and } \int_0^\infty e^{-\omega t} \, |f_m(t)| \, dt < \infty$$

Therefore, by the Parseval Identity for f_m and $\bar{f}_m$, we have that finfing the minimum of $I_m(\omega;s_1,\dots,s_n)$ in the class F_m is the same as determining the minimum of

$$\bar{I}_m(\omega;s_1,\dots,s_n) = \int_{\omega/2 - i\infty}^{\omega/2 + i\infty} |\bar{f}(s) - \bar{f}_m(s)|^2 \, ds \tag{2.6}$$

in the class $\bar{F}_m$. So, from a theoretical point of view, the problem of the numerical inversion of the Laplace Transform by means of two-point Padé Approximation consists in finding the best 2PTA to $\bar{f}$ in the L^2-norm (2.6).

With respect to the first question, namely the existence of a minimum, we need the following technical result, which can be proved after some straightforward calculations.

LEMMA 1. The application $s(z) = \omega \frac{z}{z-1}$ maps conformally the unit circle $|z|<1$ onto the half-plane of convergence Re s > $\omega/2$. So, if $\bar{f}$ admits the expansions $\bar{f}(s) = \sum_{j=0}^{\infty} c_j s^j \equiv \bar{f}_0$ and $\bar{f}(s) \sim \sum_{j=1}^{\infty} d_j s^{-j} \equiv \bar{f}_\infty$ $(s\to\infty)$, and $\bar{f}_m$ is a (k(m)/m) 2PTA to the pair $(\bar{f}_0,\bar{f}_\infty)$, we have that the function $g(z) = (1-z)^{-1}\bar{f}(s(z))$ admits the expansions

$$g(z) = \sum_{j=0}^{\infty} c_j^{*} z^j \equiv g_0 \tag{2.7}$$

$$g(z) \sim \sum_{j=1}^{\infty} d_j^*(z-1)^j \equiv g_1 \quad (s\to 1) \tag{2.8}$$

and the rational function $g_m(z)= (1-z)^{-1}\overline{f}_m(s(z))$ is a $(k(m)/m)$ 2PTA in the points 0 and 1 to g and denoted by $(k/m)_g^{(0,1)}$ (For more details about 2PTAs in two finite points see e.g.[5] and references there found).

Thus, given a function g holomorphic on the unit circle and satisfying (2.7) and (2.8), we must find the best approximation to g in the class

$$G_m = \left\{ g_m(z) = (k/m)_g^{(0,1)}(z) = \left(\sum_{j=0}^{m-1}\alpha_j z^j\right) \Big/ \left(\sum_{j=0}^{m}\gamma_j z^j\right), \text{ where } \sum_{j=0}^{m}\gamma_j z^j \neq 0 \text{ in } |z|<1 \right\}$$

with the L^2-norm: $\|g\| = \left| \int_C |g(z)|^2\, dz \right|^{1/2}$, where $C \equiv |z| = 1$.

Now, we can establish the following

THEOREM 1. Let g(z) be a fuction as above and defined on $|z| = 1$ and let r be a function in the set G_m so that $I(r) = \int_C n(z)|g(z)-r(z)|^2\, dz < \infty$, where n is a nonnegative,integrable function on C. Then, there exists $R \in G_m$ such that $I(r) \geq I(R)$, for any $r \in G_m$.

Proof

It is clear that $I(r) > 0$ for any $r \in G_m$.

Thus, if $b = \inf\{I(r)/\ r \in G_m\} < \infty$, there exists a sequence $\{r_j\} \subset G_m$ satisfying that $\lim_{j\to\infty} I(r_j) = \lim_{j\to\infty} \int_C n(z)|g(z)-r_j(z)|^2\, dz = b$. From a result by Walsh (see [6], p.348), there exists a subsequence $\{R_j\}$ of $\{r_j\}$ that converges continuously (uniformly in closed subsets) in $\bar{\mathbb{C}}$ to a limit R, with the possible exception of m points in $|z| > 1$.

So, for any $\rho < 1$, $\{R_j\}$ converges uniformly to R in $C_\rho \equiv |z| \leq \rho$ and by the continuity of the equations deduced from conditions (1.1) and (1.2), $R \in G_m$. Thus, for a sufficiently large j the function $n(z)|g(z)-R_j(z)|^2$ is bounded by an integrable function in C_ρ and we can write

$$b = \lim_{j\to\infty} \int_{C_\rho} n(z)|g(z)-R_j(z)|^2|dz| \geq \int_{C_\rho} n(z)|g(z)-R(z)|^2|dz|$$

Now, if we take a sequence of disks $C_{\rho_n} \equiv |z| = \rho_n$ (with $\rho_n \to 1$), one yields $\int_C n(z)|g(z)-R(z)|^2|dz| \leq I(r)$, for any $r \in G_m$. ∎

As an immediate consequence of Theorem 1, (2.5) and Lemma 1, we have

COROLLARY

There exists the minimum of $\{\|f-f_m\|$, where $f_m \in F_m\}$

Once we have guaranteed the existence of a minimum, it remains to study the convergence of the sequence of the $(k/m)_g^{(0,1)}$ best approximants in the norm $\left|\int_C |g(z)|^2 dz\right|^{1/2}$. In this point we need to assume the analiticity of g in the closed unit circle, which means that $\bar{f}$ is holomorphic on a closed domain D containing the point of infinity.

THEOREM 2. Under the above conditions, the sequence of best approximants to f in the class F_m $(m \in \mathbb{N})$ converges to f in the L^2-norm.

Proof

Take into account that if g_m^* is a best (k(m)/m) ATP2 in $\{0,1\}$ to g in G_m, then $\bar{f}_m^*(s) = (1-z(s))^{-1} g_m^*(z(s))$ is a best (k(m)/m) 2PTA in $\{0,\infty\}$ to $\bar{f}$ in $\bar{F}_m$. But in [7] is established the convergence of the sequence of best (k(m)/m) 2PTA in $\{0,\infty\}$ (in uniform norm) to a function being holomorphic on a closed set D containing $\{0,\infty\}$. So, the corresponding $\{R_m^*\}$ converges uniformly to g in a closed set containing the unit circle (see [5]). Therefore, given $\varepsilon>0$, $|g(z)-R_m^*(z)| < \varepsilon$ for $|z| = 1$. But since $R_m^* \in G_m$, we conclude that

$$\int_C |g(z)-g_m^*(z)|^2 dz \le \int_C |g(z)-R_m^*(z)|^2 dz < 2\pi\varepsilon^2$$

and the convergence in the L^2-norm is established. ■

3. SOME APPLICABLE RESULTS

In this section, we aim to construct easily computable sequences of Padé-type Approximants with uniform convergence to the function g in the closed unit circle $|z| = 1$. For this, suppose that g is holomorphic in a circle $|z| \le R$ $(R>1)$, satisfying (2.6) and (2.7) (now, (2.7) is a power series).

Firstly, the one-point case (k(m) = m) will be treated. In order to study the uniform convergence of sequences of 1PTAs to g, we need the following integral representation of the error of approximation, which is a particular case of the error formula for rational interpolants with fixed poles given in [6] (pp. 186-187). Indeed, for any z such that $|z| < R$, one has:

$$E_m(z) = g(z) - g_m(z) = \frac{z^m}{2\pi i P_m(z)} \int_{C_R} \frac{P_m(x)\ g(x)}{x^m\ (x-z)}\, dx \tag{3.1}$$

where $C_R = \{z \in \mathbb{C} / \ |z|=R\}$.

One of the most interesting cases, because of the simplicity, is when g is approximated by means of 1PTAs with a single pole, i.e.: $g_m(z) = \dfrac{P_{m-1}(z)}{(z-a)^m}$.
For this choice, proposed intially by V. Iseghem [8], we have for $|z| \le 1$

$$|E_m(z)| \le C \frac{|z|^m}{R^m} \frac{||P_m||_{C_R}}{|P_m(z)|}$$

(where C only depends on f and R). Therefore,

$$|E_m(z)| \le C \frac{|z|^m}{R^m} \frac{(R+|a|)^m}{(|a|-1)^m}$$

Thus, if $|a| > \frac{2R}{R-1}$, we have that $\lim_{m\to\infty} |E_m(z)|^{1/m} < 1$ and the following Theorem has been proved

THEOREM 3. If g is holomorphic on C_R (R>1), the sequence of 1PTAs with denominators given by $P_m(z) = (z-a)^m$ (where $|a|>(2R/R-1)$) converges with geometrical rate to g in the closed unit circle. Moreover, the asymptotic degree of convergence is

$$\lim_{m\to\infty} |E_m(z)|^{1/m} \le \frac{R+|a|}{R(|a|-1)}, \text{ for any z with } |z|\le 1$$

Another choice of easy computation arise when the poles are uniformly distributed on the boundary $|z| = R$, that is for example, when the poles are taken as the mth root of R^m: $P_m(z) = z^m - R^m$. In this case, by applying (3.1), we obtain

$$\lim_{m\to\infty} |E_m(z)|^{1/m} \le \frac{|z|}{R} \lim_{m\to\infty} \frac{||P_m||_{C_R}^{1/m}}{|P_m(z)|^{1/m}} \le \frac{|z|}{R} < 1$$

and therefore

THEOREM 4. If g is holomorphic in C_R (R>1), the sequence of 1PTAs with denominators $P_m(z) = z^m - R^m$ converges geometrically to g on compacts of $|z|\le R$ (in particular, on the closed unit circle). In this case, the degree of convergence on the unit circle is $\lim_{m\to\infty} |E_m(z)|^{1/m} \le \frac{1}{R}$

Now, let us consider the use of sequences of (k(m)/m) 2PTAs (in $\{0,1\}$) to g. To study the convergence of such sequences, we make use of the error formula (see e.g. [6])

$$E_{k(m)}(z) = g(z) - g_{k(m)}(z) = \frac{z^{k(m)}(z-1)^{m-k(m)}}{2\pi i\, P_m(z)} \int_{C_R} \frac{g(x)\, P_m(x)}{x^{k(m)}(x-1)^{m-k(m)}} \frac{dx}{x-z} \qquad (3.2)$$

Firstly, observe that the choice of denominators in Theorem 4 is such that $\lim_{m\to\infty}\left|\frac{Q_m(z)}{z^m}\right|^{1/m} = |\phi(z)| = |R/z|$ uniformly on any compact of C_R not containing the origin, where ϕ maps conformally the interior of C_R onto the exterior of the unit circle (so that $z = 0$ corresponds to the point of infinity). This fact suggests to make use of the Möbius transformation $z = \frac{R^2(w+1)}{R^2+w}$, which maps C_R onto itself but $z = 1$ corresponds to $w = 0$. Under these condiitions, as $w = \frac{R^2(z-1)}{R^2-z}$, we choose the sequence of denominators $P_m(z) = P_{k(m)}(z)\, P_{h(m)}(z)$, where $h(m) = m-k(m)$ and

$$P_{k(m)}(z) = z^{k(m)} - R^{k(m)} \tag{3.3}$$

$$P_{h(m)}(z) = (R(z-1))^{h(m)} - (R^2-z)^{h(m)} \tag{3.4}$$

With this choice of denominators, if $\lim_{m\to\infty}\frac{k(m)}{m} = p$ $(0\le p\le 1)$, we have by (3.2)

$$\lim_{m\to\infty}|E_k(z)|^{1/m} \le \frac{|z|^p}{R^p}|z-1|^{1-p}\lim_{m\to\infty}\frac{1}{|P_h(z)|^{1/m}}\left\{\int_{C_R}\frac{|P_h(x)|\,|g(x)|}{|x-1|^h\,|x-z|}|dx|\right\}^{1/m}$$

So, if we make the transformation: $x = \frac{R^2(t+1)}{R^2+t}$ and $z = \frac{R^2(w+1)}{R^2+w}$, taking into account (3.3) and (3.4), one yields

$$\lim_{m\to\infty}|E_k(z)|^{1/m} \le \frac{|z|^p}{R^p}\lim_{m\to\infty}\left\{\frac{|w|^h}{|w^h-R^h|}\int_{C_R}\frac{|t^h-R^h|\;|R^2+w|}{|t|^h\;|t-w|\;|R^2+t|}|dt|\right\}^{1/m} \le$$

$$\le \frac{|z|^p}{R^p}\,\frac{|w|^{1-p}}{R^{1-p}} < 1$$

Therefore, we can state

THEOREM 5. The sequence of (k(m)/m) 2PTAs in $\{0,1\}$ to g with denominators $P_m(z) = P_{k(m)}\cdot P_{h(m)}$ as above converges geometrically to g in compacts of $|z| < R$.

Remark 1

Observe that for the sequence of denominators in Theorem 5, we obtain that

$$\lim_{m\to\infty}|E_{k(m)}(z)|^{1/m} \le \frac{|z|^p}{R^p}\,\frac{R^{1-p}|z-1|^{1-p}}{|R^2-z|^{1-p}}$$

where $\psi(z) = \frac{R^2-z}{R(z-1)}$ maps conformally $|z| < R$ onto the exterior of the unit

circle, now with $\psi(1) = \infty$.
Moreover, if we denote by $g_{D_R}(z;0) = \mathrm{Log}|\phi(z)|$ and $g_{D_R}(z;1) = \mathrm{Log}|\psi(z)|$ the Green's Functions of $D_R \equiv |z| < R$ with respective poles in $z = 0$ and $z = 1$, we have that $\lim_{m\to\infty} |E_{k(m)}(z)|^{1/m} \le \exp\{-G(\mu,C_R;z)\}$, where $\mu = p\delta_0 + (1-p)\delta_\infty$ (δ_0 and δ_∞ denote the respective Dirac measures with unit mass in 0 and 1) and

$G(\mu,C_R;z) = \int g_{D_R}(z;x)\, d\mu(x)$ is the Green's Potential of the distribution μ (see e.g. Landkof [9]).

4. AN ALTERNATIVE APPROACH

Throughout this section an alternative method to get the numerical inversion of the Laplace Transform via Padé Approximation will be discussed. This approach consists of inverting term by term the expansion of $\bar{f}$ near the point of infinity to obtain an expansion of the original function around the origin. If we get a formal power series, sequences of 1PTAs can be used. For example, if this power series defines an entire function, by a straightforward application of (3.1) is easy to prove that the sequence of 1PTAs with denominators $P_m(z) = (z-a_m)^m$, where $\lim_{m\to\infty} a_m = \infty$, converges uniformly to f in compacts of $\mathbb{C}$ and hence, we get a global estimation of the original function f in $[0,\infty)$. However, even in the case of an entire function, the convergence could be slow for large values of t.

This drawback suggest the convenience of making use of 2PTAs. A result by Doetsch ([4], p.195 and 254) provide the desired expansion for f near the infinity. As an illustration of this situation, consider the example $\bar{f}(s) = s^{-1/2}(s^{1/2}+a)^{-1}$ proposed by Grundy [10]. In this case, we have that

$$\bar{f}(s) = \frac{1}{s}\sum_{n=0}^{\infty}(-1)^n (as^{-1/2})^n, \text{ for } |s^{1/2}| > a$$

$$\bar{f}(s) = \frac{1}{as^{1/2}}\sum_{n=0}^{\infty}(-1)^n (s^{1/2}/a)^n, \text{ for } |s^{1/2}| < a$$

and by applying the result by Doetsch, we get

$$f(t) \sim \sum_{n=0}^{\infty} \frac{(-1)^n\, t^{-(n+1)/2}}{a^{n+1}\,\Gamma((1-n)/2)} \quad (t\to\infty) \quad \text{and} \quad f(t) = \sum_{n=0}^{\infty} \frac{(-1)^n\, a^n\, t^{n/2}}{\Gamma(1+n/2)}$$

So, setting $t^{1/2} = z$, we obtain two expansions g and h of the form

$$g(z) = \sum_{n=0}^{\infty} c_n z^n, \text{ where g is an entire function}$$

$h(z) = \sum_{n=1}^{\infty} d_n z^{-n}$, with absolute convergence for $|z|>R'$

Now, we are concerned with the convergence of sequences of (k(m)/m) 2PTAs to the pair (f,g) with denominators of the form $P_m(z) = (z+\alpha_m)^m$ ($\alpha_m>0$). For this, we can consider that f is analytical in $|z|\le R$ (R arbitrarily large) and g is analytical in $|z|\ge R''$(where $R''>\max\{R,R'\}$) and then, by applying the integral error formula for 2PTAs in a unconnected domain (see [7]), one has

$$E_{k(m)}(z) = \frac{z^k}{2\pi i P_m(z)} \left\{ \int_{C_0} \frac{P_m(x)f(x)}{x^k(x-z)} dx + \int_{C_\infty} \frac{P_m(x)f(x)}{x^k(x-z)} dx \right\} \tag{3.5}$$

where $C_0 \equiv |z|=R$ and $C_\infty \equiv |z|=R'$ are orientated so that the origin and the infinity lie respectively in its interior. Thus, if $|z|\le R_0$ ($R_0<R$) and Re(z)>0, one yields

$$|E_{k(m)}(z)| \le \frac{e^{qR}\, M(R)}{2\pi} (R_0/R)^{k(m)} + \frac{e^{qR''}\, M(R'')}{2\pi} (R_0/R'')^{k(m)}$$

where $q = \lim_{m\to\infty} \frac{m}{\alpha(m)}$ and $M(R) = ||f||_{C_0}$, $M(R'') = ||g||_{C_\infty}$. Furthermore, if $\lim_{m\to\infty} \frac{k(m)}{m} = p$ (now $0<p\le 1$), the asymptotic degree of convergence is given by

$$\lim_{m\to\infty} |E_{k(m)}(z)|^{1/m} \le \max\left\{(R_0/R)^p\ ,\ (R_0/R'')^p\right\} = (R_0/R)^p < 1$$

These results are summarized in the following

THEOREM 6. Under the above conditions, the sequence of (k(m)/m) 2PTAs to the pair (g,h) converges geometrically in any compact subset of $[0,\infty)$.

Remark 2. We can see from Theorem 7 that an easily computable global approximation can be constructed, where the inclusion of a few terms in the expansion near the infinity provides results which improves the convergence.

REFERENCES

1. G. A. Baker and P. Graves-Morris, Padé Approximants, Part I: Basic Theory, in Encyclopedia of Mathematics and its Applications, Cambridge University Press (1.984)

2. C. Brezinski, Padé-type Approximants and General Orthogonal Polynomials, Birkauser Verlag, Basel, (1.980)

3. A. Draux, Publication A.N.O. 110 (1983)

4. G. Doetsch, Introduction to the theory and application of the Laplace transformation, Springer Verlag, Berlin (1974)

5. P. González-Vera, R. Orive, M. Jiménez and J. D. Betancor, *XIV Jornadas*

Hispanolusas (1.989), pp.

6. J. Walsh, Interpolation and Approximation by Rational Functions in the complex domain, Amer. Math. Soc., Colloq. Public., Providence, R. I. (1969)

7. P. González-Vera and R. Orive, J. Comp Appl. Math., in press

8. J. Van Iseghem, Applications des Approximants de type Padé, Ph. D. Université des Sciences and Techniques de Lille (1982)

9. N. S. Landkof, Foundations of Modern Potential Theory, Grundlehren der Matematischen Wissenschaften, 190, Springer-Verlag, New York (1972)

10. R.E.Grundy, J. Inst. Maths.Applics. 20 (1977), pp. 299-306

2nd Int. Coll. on Numerical Analysis, pp. 83-92
D. Bainov and V. Covachev (Eds)

On a class of singularly perturbed boundary value problems for which an adaptive mesh technique is necessary

P. W. Hemker and G. I. Shishkin[1]
CWI, Amsterdam, The Netherlands,
IMM, Ural Branch of the Russian Academy of Science, Ekaterinburg, Russia.

Abstract - We construct discrete approximations for linear boundary value problems for elliptic differential equations for which the coefficient of the highest derivatives can take arbitrary small values from the interval $(0, 1]$. Discretisation errors for classical discrete methods depend on the value of this parameter and can be of a size comparable with the solution of the original problem. We describe how to construct special discrete methods for which the accuracy of the discrete solution does not depend on the value of the parameter, and only depends on the number of mesh points used. We describe a class of boundary value problems, for which an adaptive mesh technique is necessary to obtain a parameter-uniform error-estimate as mentioned above. For this class of problems a special finite difference method is constructed. The solutions, obtained by the special method, converge in the discrete l^∞-norm. For a model problem we show and compare results obtained by the classical method and those for the special scheme.

1 Introduction

For the solution of boundary value problems that have a smooth solution, well established methods can be used, such as finite difference, finite element or finite volume schemes (see e.g. [5]). The accuracy of the approximate solutions deteriorates when the solution becomes less smooth.

In the case of singularly perturbed differential equations, when the coefficient of the highest derivatives can take an arbitrary small value in $(0, 1]$, the solution of the boundary value problems may have limited smoothness. The derivatives of such solutions may increase without bound when the small parameter tends to zero. When classical numerical methods are used, the error of the approximate solution depends on the value of the small parameter and may be of a size comparable with the solution of the original problem [4]. Therefore, for singularly perturbed boundary value problems it makes sense to construct special numerical methods, for which accuracy of approximate solution does not depend on the parameter, and for which the size of the error depends only on the number of nodal point used, i.e. methods which converge uniformly with respect to the small parameter (parameter-uniformly).

In [2, 4] special difference schemes were constructed for the solution of singularly perturbed boundary value problems, and it was proved that these schemes are parameter-

[1]This research was supported in part by the Dutch Research Organisation NWO through grant 07-30-012 .

uniformly convergent. Those first strong results belong to two different approaches for the construction of special numerical methods for these problems with boundary layers:

(a) In the "fitted" method described in [4], the coefficients of difference equations are adapted such that a parameter-uniform accuracy of the approximate solution can be guaranteed on meshes with an arbitrary distribution of nodes (for example, on a uniform mesh);

(b) In the "adaptive mesh" method, described in [2], the usual classical discretisation scheme is used, but the nodes in the mesh are redistributed (condensed in the boundary layer) such that the parameter-uniform convergence is achieved.

The methods from the first approach are attractive because they allow the use of meshes with an arbitrary distribution of nodes, in particularly uniform grids (see, for example, [1, 4]). Methods based on the second approach were constructed for a series of boundary value problems, e.g. see [8] (and references therein). For some boundary value problems, parameter-uniformly convergent schemes were constructed using both approaches for the same problem ([8]), or with the different approaches used to approximate the derivatives in different coordinate directions [7].

In this paper we consider a class of singularly perturbed boundary value problems for elliptic equations. Using the approaches (a) and (b), in Section 3 we introduce a natural class B of finite difference schemes for boundary value problems, in which we look for the special schemes which yield a parameter-uniformly convergent approximation to the solution.

The considered class of singularly perturbed convection-diffusion problems describe the diffusion of some material in a moving medium. For these boundary value problems we construct a special parameter-uniformly convergent schemes. We show that for schemes from class B the use of special adapted meshes is necessary. With a special adapted mesh we are able to construct finite difference schemes which converge parameter-uniformly. In Section 5 we give numerical results and compare the classical and the special finite difference schemes.

2 The class of boundary value problems studied

In many problems from physics we recognise the convective transport with diffusion described by

$$-\varepsilon\Delta u(x) + <\vec{v}(x)\cdot\vec{\nabla}> u(x) = F(x), \ x \in \Omega. \tag{2.1a}$$

For the boundary of Ω, which we assume to be piecewise smooth, we distinguish a wall, $\partial\Omega^0$, an inflow and an outflow boundary, $\partial\Omega^+$ and $\partial\Omega^-$ respectively. Clearly we have $\partial\Omega = \partial\Omega^0 \cup \partial\Omega^+ \cup \partial\Omega^-$ and $<\vec{n}(x)\cdot\vec{v}(x)> = 0, x \in \partial\Omega^0$; $<\vec{n}(x)\cdot\vec{v}(x)> < 0, x \in \partial\Omega^+$; $<\vec{n}(x)\cdot\vec{v}(x)> > 0, x \in \partial\Omega^-$. Here $\vec{n}(x)$ is the outward normal unit vector. At a wall, $\partial\Omega^0$, the boundary condition is written as

$$\nu(u(x) - U(x)) + \frac{\partial}{\partial n}u(x) = 0, \quad x \in \partial\Omega^0 . \tag{2.1b}$$

For $\nu \to \infty$ condition (2.1b) reduces to the Dirichlet condition. At $\partial\Omega^+$ we assume the concentration to be known and at $\partial\Omega^-$ we assume its diffusive component to be known:

$$u(x) = U^+(x), \quad x \in \partial\Omega^+ \tag{2.1c}$$

$$\frac{\partial}{\partial n}u(x) = \Psi(x), \ x \in \partial\Omega^-. \tag{2.1d}$$

When the parameter ε tends to zero, we may expect a boundary layer to appear in the neighbourhood of the boundary $\partial\Omega^0$.

In this paper we describe a class of two-dimensional diffusive transport problems, for which we construct our special finite difference schemes. The problem is slightly more general than problem (2.1). On the rectangular domain $\Omega = \{x:\ 0 < x_i < d_i,\ i = 1,2\}$ we consider the elliptic boundary value problem for equation

$$L_{(2.2)}u(x) \equiv \{\varepsilon^2 \textstyle\sum_{s=1,2} a_s(x)\frac{\partial^2}{\partial x_s^2} - b(x)\frac{\partial}{\partial x_1} - c(x)\}u(x) = f(x), \quad x \in \Omega, \tag{2.2a}$$

$$u(x) = \varphi(x), \quad x \in \partial\Omega^+ = \{x \mid x \in \partial\Omega,\ x_1 = 0\}, \tag{2.2b}$$

$$\begin{aligned} l_{(2.2)}u(x) \equiv\ & -\alpha\varepsilon\tfrac{\partial}{\partial n}u(x) - (1-\alpha)u(x) = \psi(x), \\ & x \in \partial\Omega^0 = \{x \mid x \in \partial\Omega,\ 0 < x_1 < d_1\}, \end{aligned} \tag{2.2c}$$

$$\frac{\partial}{\partial n}u(x) = \eta(x),\ x \in \partial\Omega^- = \{x \mid x \in \partial\Omega,\ x_1 = d_1\}. \tag{2.2d}$$

Here $a_s(x),\ b(x),\ c(x),\ f(x),\ x \in \overline{\Omega};\ \varphi(x),\ x \in \partial\Omega^+;\ \psi(x),\ x \in \partial\Omega^0;\ \eta(x),\ x \in \partial\Omega^-$ are sufficiently smooth and bounded functions, which satisfy

$$0 < a_0 \le a_1(x),\ a_2(x) \le a^0, \quad 0 < b_0 \le b(x), \quad c(x) \ge 0,\ x \in \overline{\Omega}. \tag{2.2e}$$

The parameters α and ε may take any value from the intervals $\varepsilon\in(0,1]$, $\alpha\in[0,1]$. When the parameter ε tends to zero in (2.2), in the neighbourhood of the boundary $\partial\Omega^0$ parabolic boundary layers may appear. We suppose that for the functions $\phi(x)$ and $\psi(x)$ compatibility conditions are satisfies to guarantee a continuous solution and derivative with respect to x_2 on the set $\partial\Omega^0 \cap \partial\Omega^+$.

The above class of problems includes, for example, the boundary value problem for the regular differential equation:

$$L_{(2.3)}U(y) \equiv \{\sum_{s=1,2} A_s(y)\frac{\partial^2}{\partial y_s^2} - B(y)\frac{\partial}{\partial y_1}\}U(y) = F(y), \tag{2.3a}$$

for $y \in \tilde{\Omega}$, with the regular boundary conditions

$$\begin{aligned} U(y) &= \Phi(y), & y \in \tilde{\partial}\Omega^+, \\ l_{(2.3)}U(x) \equiv -\alpha\tfrac{\partial}{\partial n}U(y) - (1-\alpha)U(y) &= \Psi(y), & y \in \tilde{\partial}\Omega^0, \\ \tfrac{\partial}{\partial n}U(y) &= 0, & y \in \tilde{\partial}\Omega^-, \end{aligned} \tag{2.3b}$$

on the rectangular domain $\tilde{\Omega} = \{y \mid 0 < y_i < \tilde{d}_i,\ i = 1,2\}$, $\tilde{d}_i = \varepsilon^{-1}d_i$.

As we shall see in the next section, for boundary value problem (2.2) classical finite difference approximations on a uniform grid [5] converge to the solution of (2.2) only for a fixed value of the parameter ε (see Theorem 2.1). The accuracy of the numerical solution depends on this value of the parameter and it decreases with ε until all accuracy is lost when the value of ε/b_0 is comparable with the mesh-size of the uniform grid. We shall show that classical finite difference schemes do not converge ε-uniformly (see Theorem 2.2) for small values of the parameter ε. Therefore it is our concern to construct special schemes which do converge ε-uniformly. In [7] a particular ε-uniformly convergent finite difference scheme was constructed for the special case of a Dirichlet problem for equation (2.2a).

To solve of boundary value problem (2.2) we first study the classical finite difference method on a (possibly) non-uniform grid. On the set $\overline{\Omega}$ we introduce the rectangular grid

$$\overline{\Omega}_h = \overline{\omega}_1 \times \overline{\omega}_2, \tag{2.4}$$

where $\overline{\omega}_s$ is the (possibly) non-uniform grid on the interval $[0, d_s]$; N_s is the number of nodes in grid $\overline{\omega}_s$, $s = 1, 2$. Define $h_s^i = x_s^{i+1} - x_s^i$, $h_s = \max_i h_s^i$, $h = \max_s h_s$, $N = \min_s N_s$, $s = 1, 2$, $\Omega_h = \Omega \cap \overline{\Omega}_h$, $\partial\Omega_h = \partial\Omega \cap \overline{\Omega}_h$. For problem (2.2) we use difference scheme

$$\Lambda_{(2.5)} z(x) = f(x), \; x \in \Omega_h, \tag{2.5a}$$

$$z(x) = \varphi(x), \; x \in \partial\Omega_h^+, \tag{2.5b}$$

$$\lambda_{(2.5)} z(x) = \psi(x), \; x \in \partial\Omega_h^0, \tag{2.5c}$$

$$\delta_{\overline{x_1}} z(x) = \eta(x), \; x \in \partial\Omega_h^-, \tag{2.5d}$$

Here

$$\Lambda_{(2.5)} z(x) \equiv \varepsilon^2 \sum_{s=1,2} a_s(x) \delta_{\overline{x_s}\widehat{x}_s} z(x) - b(x) \delta_{\overline{x_1}} z(x) - c(x) z(x),$$

$$\lambda_{(2.5)} z(x) \equiv \begin{cases} +\varepsilon\alpha\delta_{x_2} z(x) & -(1-\alpha) z(x), \quad x_2 = 0, \\ -\varepsilon\alpha\delta_{\overline{x_2}} z(x) & -(1-\alpha) z(x), \quad x_2 = d_2, \end{cases}$$

$\delta_{x_s} z(x)$ and $\delta_{\overline{x_s}} z(x)$ are the forward and backward difference and $\delta_{\overline{x_s}\widehat{x}_s} z(x)$ is the second difference on the non-uniform grid.

The difference scheme (2.4)-(2.5) is monotone (i.e. the maximum principle holds) [5].

By means of the maximum principle, and taking into account estimates of the derivatives, we find that the solution of the scheme (2.4)-(2.5) converges (for a fixed value of the parameter ε) as

$$|u(x) - z(x)| \leq M\varepsilon^{-4} N^{-1}, \quad x \in \overline{\Omega}_h . \tag{2.6}$$

Here and in the following we denote by M (or m) a sufficiently large (or small) positive constants which do not depend on the value of parameter ε or on the difference operators. The proof of (2.6) is similar to the classical proof of convergence for monotone difference schemes [5, 8]. Taking into account an a-priori estimate for the solution this results in the following theorem.

Theorem 2.1 Let all k-th derivatives of the solution of boundary value problem (2.2) be bounded by $\mathcal{O}(\varepsilon^{-2k})$. Then, for a fixed value of parameter ε, the solution of the scheme (2.5)-(2.4) converges to solution of boundary value problem (2.2) with an error bound given by (2.6).

Clearly, from (2.6) no ε-uniform convergence of difference solution follows. In a report [3] we prove that it is impossible to get such an estimate for (2.4)-(2.5) on a fixed, ε-independent mesh. The proof follows the same lines as given in [6]: Thus we arrive at the following theorem.

Theorem 2.2 On a uniform grid the solution of the classical finite difference scheme (2.5) does not converge ε-uniformly to the solution of the boundary value problem (2.2).

3 Fitted difference scheme

In this section we show that there does not exist a difference scheme, even based on a fitted method, for which the solution on a uniform grid converges ε-uniformly to the solution of boundary value problem (2.2). Let us consider the boundary value problem

$$L_{(3.7)}u(x) \equiv \varepsilon^2 \Delta u(x) - \frac{\partial}{\partial x_1} u(x_1) = 0, \quad x \in \Omega, \tag{3.7a}$$

$$u(x) = 0, \quad x \in \partial\Omega^+, \tag{3.7b}$$

$$l_{(2.2)}u(x) = \psi(x), \quad x \in \partial\Omega^0, \tag{3.7c}$$

$$\frac{\partial}{\partial n} u(x) = 0, \quad x \in \partial\Omega^-, \tag{3.7d}$$

where

$$\psi(x) = \begin{cases} \psi_0(x_1), & x \in \partial\Omega^0, \quad x_2 = 0, \\ 0, & x \in \partial\Omega^0, \quad x_2 \neq 0. \end{cases} \tag{3.7e}$$

We assume that $\psi_0(0) = 0$ and that $\psi_0(x_1)$, $x_1 \in [0, d_1]$ is sufficiently smooth.

For the solution of problem (3.7) a boundary layer appears near the wall for $x_2 = 0$. We know that the main term of asymptotic expansion of this solution is the solution of a parabolic boundary value problem, with solution u_{par}, and we have the estimate

$$|u_{(3.7)}(x) - u_{\text{par}}(x)| \leq M\,\varepsilon, \quad x \in \overline{\Omega}.$$

Now we introduce the class A of finite difference schemes for the model problem (3.7). It is defined by the use of a uniform mesh $\overline{\Omega}_h^{\text{unif}}$, and the usual, fitted five-point finite difference scheme for the approximation of equation (3.7a):

$$\Lambda_{(3.8)}z(x) \equiv \{ \sum_{s=1,2} (A_s \delta_{x_s \overline{x}_s} + B_s \delta_{x_s} - C\} z(x) = E\,. \qquad x \in \Omega_h. \tag{3.8}$$

Here the coefficients A_s, B_s, C, E are functionals derived from the coefficients of equation (3.7a), depending on x, h_1, h_2 and ε. We suppose that for $h_2\,\varepsilon^{-1} \to 0$ and $h_1 \to 0$ the coefficients A_s, B_s, C, E approximate (in ℓ^∞-norm) the coefficients in equation (3.7a), at least in the neighbourhood of a point near the boundary layer. We can prove (see [8]) the following statement.

Theorem 3.1 For the finite difference schemes in class A (i.e. on a uniform grid $\overline{\Omega}_h^{\text{unif}}$, there does not exist a discrete solution which converges ε -uniformly to the solution of boundary value problem (3.7).

A statement similar to Theorem 3.1 holds if we construct a difference scheme based on a stencil with an arbitrary, finite number of nodes. Theorem 3.1 and its generalisation can be understood as follows. All solutions of problem (3.7), for arbitrary functions $\psi(x)$ in (3.7e), are singular solutions. Even in the neighbourhood of the parabolic boundary layer, these solutions can not be represented as a linear combination of a finite set of functions of boundary layer type. Therefore we cannot choose a finite number of coefficients which define a finite difference operator such that it approximates the differential operator for the set of singular solutions. Thus, for a given (ε-independent) grid we cannot expect to find a method that is ε-uniformly convergent for problem (2.2).

Let us now introduce the class B of finite difference schemes for problem (2.2). It is defined by the use of (non-uniform) rectangular grids $\overline{\Omega}_{h\,(2.4)}$ and, for the approximation of equation (2.2a), a (fitted) five-point finite difference stencil. The coefficients of the difference operator are functionals of the coefficients of equation (2.2a) and they depend on x, ε and also on the mesh. We again suppose that for $h \to 0$ the coefficients of the difference operator approximate (in the ℓ^∞-norm) coefficients in equation (2.2a) on the set $\overline{\Omega}_{h\,(2.4)}$.

We remark that the class B is a natural class of finite difference schemes. This class includes methods both with fitted schemes and with fitted grids, i.e. it allows all freedom of choice mentioned in the introduction. The following statements can be derived from Theorem 3.1

Theorem 3.2 For the boundary value problem (2.2), there does not exist a finite difference scheme in class B, based on a fitted scheme, for which the solution converges ε -uniformly to solution of the boundary value problem (2.2) for a mesh with an arbitrary ε-independent distribution of nodes.

Theorem 3.3 For the boundary value problem (2.2), for schemes in class B special ε-dependent grids (adaptive meshes) are necessary for the construction of ε -uniformly convergent schemes.

Hence, we should conclude that, if we consider problems with parabolic boundary layers, within class B adaptive mesh techniques are necessary for the construction of ε-uniformly convergent difference schemes. Such an adaptive method should provide an (a-priori) ε-dependent mesh, or it should adapt its mesh during the computation, using either data from the problem (the given value of ε) or from the solution.

4 The ε-uniformly convergent difference scheme

By means of an adapted condensed mesh, in this section we construct an ε-uniformly convergent method for boundary value problem (2.2). From the previous section it is clear that we need a special mesh, condensed in the neighbourhood of boundary layers. How the nodes have to be placed is derived from a-priori estimates of the solution and its derivatives. Let us consider the special grid (see [8]),

$$\overline{\Omega}_h^* = \overline{\omega}_1 \times \overline{\omega}_2^* \,, \tag{4.9}$$

where $\overline{\omega}_1$ is a uniform mesh, and $\overline{\omega}_2^* = \overline{\omega}_2^*(\sigma)$ is a special piecewise uniform mesh; σ is a parameter depending on ε and N_2. The mesh $\overline{\omega}_2^*(\sigma)$ is constructed as follows. The interval $[0, d_2]$ is divided in three parts $[0,\sigma]$, $[\sigma, d_2-\sigma]$ and $[d_2-\sigma, d_2]$, $0 < \sigma \le d_2/4$. We divide each interval $[0,\sigma]$ and $[d_2-\sigma, d_2]$ in $N_2/4$ equal parts and interval $[\sigma, d_2-\sigma]$ in $N_2/2$ equal parts. We assume that $\sigma = \sigma(\varepsilon, N_2) = \min[d_2/4,\ m\varepsilon \ln N_2]$, where $m = m_{(4.9)}$ is an arbitrary positive number.

Theorem 4.1 If the solution of (2.2) satisfies some estimates described in [3], then the solution of the discrete problem (2.5)-(4.9) converges ε-uniformly to the solution of boundary value problem (2.2). In particular, for the solution of the discrete problem the estimate holds:

$$|u(x) - z(x)| \le M\,N^{-1/3}\,, \quad x \in \overline{\Omega}_h^* \,. \tag{4.10}$$

The proof of this theorem follows the lines of the proof in [8].

The difference scheme (2.5)-(4.9) belongs to class B. This scheme has typically an a-priori adapted mesh.

5 Numerical results

By the theory (see Theorem 2.2) it is shown that the classical difference scheme (2.5) on the uniform grid does not converge in the l^∞-norm ε-uniformly to the solution of the boundary value problem (2.2). But it would be possible that the error $\max_{\overline{\Omega}_h} |z(x)-u(x)|$ is not too large for the classical scheme. That would reduce the need for a special scheme.

On the other hand, Theorem 4.1 shows that the special scheme (2.5),(4.9) converges ε-uniformly, but no indication is given about the value of the order constant M in (4.10) and the order of convergence is rather small. It might be possible that the error is rather large for any reasonable value of N_1, N_2. This might reduce the practical value of the special scheme. To decide on the practical value of the new scheme numerical experiments should give the final answer.

To see the effect of the special scheme in practice, we study the singularly perturbed elliptic equation with a mixed boundary condition.

$$L_{(5.11)}u(x) \equiv \varepsilon^2 \Delta u(x) - \frac{\partial}{\partial x_1} u(x) = -1, \quad x \in \Omega \;, \tag{5.11}$$

$$l_{(5.11)}u(x) = \psi(x), \quad x \in \partial\Omega^0 \,, \quad u(x) = 0, \quad x \in \partial\Omega^+ \,, \qquad \frac{\partial}{\partial x_2} u(x) = 0 \,, \quad x \in \partial\Omega^- \,,$$

where

$$l_{(5.11)}u(x) \equiv \begin{cases} \alpha\varepsilon \frac{\partial}{\partial x_2} u(x) - (1-\alpha)u(x), & x_2 = 0 \,, \\ -\alpha\varepsilon \frac{\partial}{\partial x_2} u(x) - (1-\alpha)u(x), & x_2 = 1 \,. \end{cases}$$

We compare the numerical results for the scheme (2.5) on the uniform and on the special grid (4.9). Here $\Omega = \{x : 0 < x_1, x_2 < 1\}$,

$$\psi(x) = \begin{cases} x_1, & x \in \partial\Omega^0, x_2 = 0 \,, \\ 0, & x \in \partial\Omega^0, x_2 = 1 \,. \end{cases}$$

For the solution of problem (5.11) we have the representation

$$u(x) = U(x) + W(x), \quad x \in \overline{\Omega} \,,$$

where $U(x) = x_1$, $x \in \overline{\Omega}$, is the outer solution, and $W(x)$ represents the parabolic boundary layer in the neighbourhood of the edges at $x_2 = 0$ and $x_2 = 1$. For the solution we have the estimate

$$-1 \leq u(x) \leq 1, \quad x \in \overline{\Omega} \,.$$

Due to Theorem 4.1 the solution of the discrete problem with the adapted mesh converges ε-uniformly to the solution of our model problem (5.11). The function $u_h(x)$, which is the solution of the special scheme (2.5)-(4.9) is shown in Figure 1.

To see the difference between the use of the uniform and the adapted grid, for the approximation of (5.11) we first use the uniform grid. We solve the problem for different values of the meshwidth $h_1 = h_2 = N^{-1}$ and for different values of the parameters ε and α. The results for a set of numerical experiments is given in Table 1. From Table 1 we can see that this solution does not converge ε-uniformly. For a fixed value of N, the errors depend on the parameters ε and α. For $\varepsilon \geq 0.1$ and $\alpha = 0.0, 0.1, 0.5, 1.0$ the error behaviour is regular: when N increases, the error decreases. For $\varepsilon = 10^{-3}$ and $\alpha = 0.0, 0.1, 0.5$ and for $\varepsilon = 10^{-2}$ and $\alpha = 0.0$, for some values of N the error increases with increasing N. For $\alpha = 0.5, 1.0$ and a fixed N the error increases with decreasing ε. In particular, for $\varepsilon = 10^{-3}$ and $\alpha = 0.5, 1.0$ the errors for $N \leq 128$ are of the same

Figure 1. Solution computed with the adapted mesh:
$\varepsilon = 0.01$, $\alpha = 0.5$, $N = 32$ and $m = 1.0$.

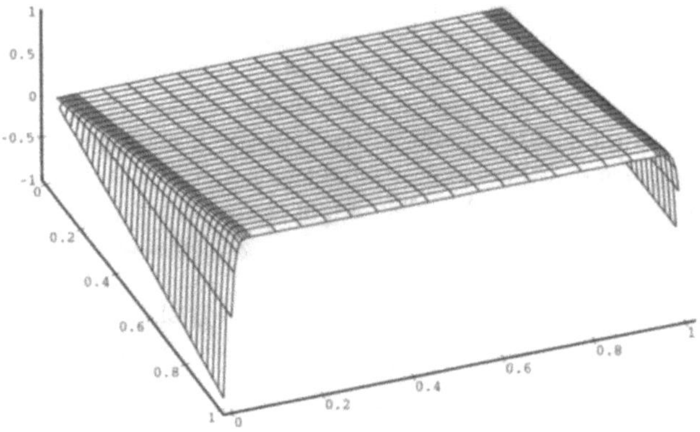

order or larger (in ℓ^∞-norm) than the solution of the BVP. Thus, the numerical results illustrate that the lack of ε-uniform convergence leads to large errors indeed.

We show the behaviour of (2.5), (4.9), with $m = m_{(4.9)} = 1$, applied to the model problem (5.11) The results for a set of numerical experiments is given in Table 2. We solve the problem for different values of N, $N = N_1 = N_2$, and for different values of the parameters ε and α. From Table 2 we can see that the solution of the scheme (2.5)-(4.9) does converge ε-uniformly indeed. The errors for a fixed value of $\varepsilon = 1.0, 10^{-1}, 10^{-2}, 10^{-3}$ and $\alpha = 0.0, 0.1, 0.5, 1.0$ have all a regular behaviour and decrease for increasing N. For a fixed value of α and N the error stabilises for decreasing ε: the errors for $\varepsilon = 10^{-2}$ and $\varepsilon = 10^{-3}$ are practically the same. For $\varepsilon \le 10^{-2}$ and a fixed value of N we find the largest error for $\alpha = 1.0$. In particular, for $\varepsilon \le 10^{-2}$, $\alpha = 1.0$ and $N = 128$ the error is less than 6%. Also here, the numerical results illustrate the practical value of ε-convergent methods.

Conclusion

For the elliptic boundary value problem (2.2), where a small parameter multiplies the highest derivative, we analysed different approaches for the construction of discrete methods. We studied methods for which the accuracy of the discrete solution does not depend on the value of the small parameter, but only on the number of points in the discretisation (ε-uniform methods extended to non-uniform grids).

We show that in a natural class of finite difference schemes no ε-uniform methods exist on a uniform grid for the problem considered (Theorem 3.1). As a consequence, for the construction of ε-uniform methods the use of an adapted non-uniform mesh is necessary. With a special, adapted, non-uniform mesh and a simple classical difference scheme, we are able to construct an ε-uniform approximation

To illustrate the practical importance of our study, we show by a numerical example that, on a uniform grid, the classical difference scheme is inaccurate (and not ε-uniformly convergent). In our example, the error (with a Neumann boundary condition) is not less than 700% of the solution, for N=128 and $\varepsilon = 10^{-3}$. The same example shows that we obtain an accurate, ε-uniformly convergent solution if we use the adapted mesh. Now

Table 1. Table of errors $E(N, \varepsilon, \alpha)$ for the classical scheme.
In this table the error $E(N, \varepsilon, \alpha)$ is defined by

$$E(N, \varepsilon, \alpha) = \max_{x \in \overline{\Omega}_h} |e(x; N, \varepsilon, \alpha)| , \tag{5.12a}$$

$$e(x; N, \varepsilon, \alpha) = z(x) - u^{\star}(x) , \tag{5.12b}$$

where $u^{\star}(x)$ is the accurate interpolation of $z_m^{\star 256}(x)$, $m = m_{(4.9)} = 1$ (see Table 2), and $z(x) \equiv z^N(x)$ is the solution of (2.5), on the uniform grid with $h_1 = h_2 = N^{-1}$. Notice that $u^{\star}(x)$ is an accurate approximation of $u(x)$.

N		4	8	16	32	64	128
ε	α						
1	0.00	0.105	0.633(-1)	0.329(-1)	0.158(-1)	0.696(-2)	0.241(-2)
10^{-1}		0.305	0.144	0.655(-1)	0.291(-1)	0.122(-1)	0.403(-2)
10^{-2}		0.247	0.127	0.822(-1)	0.107	0.882(-1)	0.246(-1)
10^{-3}		0.246	0.121	0.588(-1)	0.283(-1)	0.157(-1)	0.200(-1)
1	0.10	0.691(-1)	0.381(-1)	0.170(-1)	0.744(-2)	0.314(-2)	0.104(-2)
10^{-1}		0.312	0.149	0.701(-1)	0.310(-1)	0.129(-1)	0.423(-2)
10^{-2}		0.247	0.194	0.185	0.156	0.109	0.583(-1)
10^{-3}		0.246	0.209	0.216	0.216	0.211	0.197
1	0.5	0.852(-1)	0.401(-1)	0.183(-1)	0.842(-2)	0.359(-2)	0.119(-2)
10^{-1}		0.706	0.473	0.254	0.121	0.526(-1)	0.176(-1)
10^{-2}		1.20	1.26	1.15	0.893	0.557	0.279
10^{-3}		1.28	1.43	1.49	1.48	1.41	1.27
1	1.0	0.123	0.610(-1)	0.288(-1)	0.132(-1)	0.562(-2)	0.187(-2)
10^{-1}		1.52	0.752	0.344	0.154	0.643(-1)	0.211(-1)
10^{-2}		18.2	10.3	5.19	2.40	1.02	0.408
10^{-3}		187.	109.	57.9	29.5	14.6	7.01

the error is not larger than 6% of the solution, for any value of the parameter ε. Thus, the numerical example illustrates that the theoretical considerations have significant practical implications.

References

[1] D.N. Allen and R.V. Southwell. Relaxation method applied to determine the motion, in two dimensions, of viscous fluid past a fixed cylinder. *Quart. J. Mech. Appl. Math.*, 8:129–145, 1955.

[2] N.S. Bakhvalov. On optimization of methods to solve boundary value problems with boundary layers. *Zh. Vychisl. Mat. Mat. Fiz.*, 9:841–859, 1969. (in Russian).

[3] P. W. Hemker and G. I. Shishkin. Singularly perturbed boundary value problems for which an adaptive mesh technique is necessary. 1993, to appear.

[4] A.M. Il'in. Difference scheme for a differential equation with a small parameter at the highest-order. *Mat. Zametki*, 6:237–248, 1969. (in Russian).

[5] A.A. Samarsky. *Theory of Difference Scheme.* Nauka, Moscow, 1989. (in Russian).

Table 2. Table of errors $E(N, \varepsilon, \alpha)$ for the special scheme.
In this table the function $E(N, \varepsilon, \alpha)$ is defined by (5.12), but now $z(x) = z_m^{\star N}(x)$ in (5.12) is the solution of (2.5),(4.9) with $m = m_{(4.9)} = 1$.

N		4	8	16	32	64	128
ε	α						
1	0.0	0.105	0.633(-1)	0.329(-1)	0.158(-1)	0.696(-2)	0.241(-2)
10^{-1}		0.262	0.144	0.655(-1)	0.291(-1)	0.122(-1)	0.403(-2)
10^{-2}		0.246	0.147	0.807(-1)	0.361(-1)	0.148(-1)	0.497(-2)
10^{-3}		0.246	0.147	0.807(-1)	0.361(-1)	0.148(-1)	0.497(-2)
1	0.1	0.691(-1)	0.381(-1)	0.170(-1)	0.744(-2)	0.314(-2)	0.104(-2)
10^{-1}		0.276	0.149	0.701(-1)	0.310(-1)	0.129(-1)	0.423(-2)
10^{-2}		0.246	0.170	0.887(-1)	0.461(-1)	0.240(-1)	0.924(-2)
10^{-3}		0.246	0.169	0.887(-1)	0.461(-1)	0.241(-1)	0.925(-2)
1	0.5	0.852(-1)	0.401(-1)	0.183(-1)	0.842(-2)	0.359(-2)	0.119(-2)
10^{-1}		0.611.	0.473	0.254	0.121	0.526(-1)	0.176(-1)
10^{-2}		0.539	0.511	0.361	0.217	0.111	0.420(-1)
10^{-3}		0.535	0.511	0.361	0.217	0.111	0.420(-1)
1	1.0	0.123	0.610(-1)	0.288(-1)	0.132(-1)	0.562(-2)	0.187(-2)
10^{-1}		1.14	0.752	0.344	0.154	0.643(-1)	0.211(-1)
10^{-2}		0.977	0.889	0.554	0.301	0.144	0.521(-1)
10^{-3}		0.963	0.888	0.554	0.301	0.144	0.521(-1)

[6] G.I. Shishkin. An increase in accuracy of solutions to difference scheme for parabolic equations with small parameters at highest-order derivatives. *Zh. Vychisl. Mat. Mat. Fiz.*, 24:864–875, 1984. (in Russian).

[7] G.I. Shishkin. Approximation of solutions to singularly perturbed boundary value problems with corner boundary layers. *Zh. Vychisl. Mat. Mat. Fiz.*, 27:1360–1374, 1987. (in Russian).

[8] G.I. Shishkin. Approximation of solutions to singularly perturbed boundary value problems with parabolic boundary layers. *Zh. Vychisl. Mat. Mat. Fiz.*, 29:963–977, 1989. (in Russian).

2nd Int. Coll. on Numerical Analysis, pp. 93-102
D. Bainov and V. Covachev (Eds)

Some Robust Methods for Nonlinear Parameter Estimation

D HERMEY AND G A WATSON

Department of Mathematics and Computer Science
University of Dundee
Dundee DD14HN
Scotland

Abstract This paper is concerned with numerical methods for robust nonlinear data fitting. In particular we consider the use of the l_1 norm, and also the Huber M-estimator. The case when there are errors in all the variables of the problem is given special attention.
Keywords: Nonlinear data fitting, robustness, l_1 norm, Huber M-estimator, errors-in-variables.

1. Introduction

Let observations (x_i, y_i), $i=1,..,m$ be given, where $x_i \in I\!R$, $y_i \in I\!R$, $i=1,..,m$, and assume that an appropriate model for the data is given by

$$y = f(x, \mathbf{a}), \tag{1.1}$$

where $\mathbf{a} \in I\!R^n$ is a vector of parameters. Usually the value of m greatly exceeds that of n, so that it is not possible to fit the model exactly. The traditional approach, therefore, is to define errors r_i by

$$r_i(\mathbf{a}) = f(x_i, \mathbf{a}) - y_i, \quad i=1,...,m, \tag{1.2}$$

or in vector form

$$\mathbf{r(a)} = \mathbf{f(a)} - \mathbf{y},$$

and to seek a solving the problem

$$\text{find } \mathbf{a} \in I\!R^n \text{ to minimize } \psi(\mathbf{r(a)}), \tag{1.3}$$

where $\psi: I\!R^m \to I\!R$ is a suitably chosen loss function. Normally ψ is convex, although it need not be differentiable. However, provided only that f is a differentiable function of a, then a necessary condition for a to solve (1.3) is that there exists $\mathbf{v} \in \partial\psi(\mathbf{r(a)})$ such that

$$A(\mathbf{a})^T \mathbf{v} = 0,$$

where $A(\mathbf{a})$ is the $m \times n$ matrix with (i,j) element $\partial f_i / \partial a_j$, and $\partial\psi(\mathbf{r(a)})$ denotes the subdifferential of ψ at $\mathbf{r(a)}$. This is a consequence of standard convex analysis. A point satisfying this condition is usually referred to as a stationary point, and will normally be at best a local solution of (1.3) unless f is an affine function, when the condition is also sufficient for a solution.

Of interest here are problems where f is nonlinear. Then the usual approach to the calculation of stationary points is an iterative one, where **a** is updated to a new value giving a lower value of ψ based on a truncation of the Taylor expansion

$$\mathbf{f}(\mathbf{a}+\mathbf{h}) = \mathbf{f}(\mathbf{a}) + A(\mathbf{a})\mathbf{h} + O(\|\mathbf{h}\|^2).$$

It follows that

$$\mathbf{r}(\mathbf{a}+\mathbf{h}) = \mathbf{r}(\mathbf{a}) + A(\mathbf{a})\mathbf{h} + O(\|\mathbf{h}\|^2), \tag{1.4}$$

and so **h** can be estimated by minimizing $\psi(\mathbf{r} + A\mathbf{h})$ with respect to **h**, where the dependence of quantities on **a** is not shown explicitly. In fact most methods for solving (1.3) can be interpreted as being based on the solution of a sequence of subproblems of the following form (see for instance [1] and the references given there):

$$\text{minimize } \psi(\mathbf{r} + A\mathbf{h}) + \tfrac{1}{2}\mathbf{h}^T B\mathbf{h} \tag{1.5}$$

$$\text{subject to } \|\mathbf{h}\| \le \tau,$$

where B is a given $n \times n$ symmetric matrix, τ is a suitably chosen parameter, which (like B) may change from iteration to iteration, and the norm is a given norm on $I\!R^n$. If **h** is a solution, then **a** is updated to $\mathbf{a}+\gamma\mathbf{h}$, where γ is a suitably chosen constant. For example the Gauss-Newton method corresponds to the choice $B=0$, $\tau=\infty$ at each iteration, with γ chosen to minimize or approximately minimize ψ in the direction **h** at **a**. This can perform well, but **h** may grow so that the left hand side of (1.4) is not well represented by the truncated expansion. The situation can be improved by using a subproblem of Levenberg-Marquardt or trust region type. Here again $B=0$, but τ is adjusted systematically so that γ can be set to either 0 or 1: the former case results in a reduction in τ at the next step. Under certain conditions, these methods can converge at a second order rate. However, these conditions often do not hold, and in order for superlinear or second order convergence to be obtained in very general circumstances, it is necessary to incorporate second derivative information; this is done through a suitable choice of the matrix B. This local rate of convergence control, together with the use of τ to control global convergence, enables good performance to be obtained for wide classes of problems.

Perhaps the best known example of a loss function is given by

$$\psi(\mathbf{r}) = \sum_{i=1}^{m} r_i^2,$$

corresponding to the use of the least squares norm. This can be justified statistically if the errors r_i are independent and normally distributed. However, this is usually not verifiable and is often not the case, and large errors in even one data point can seriously degrade the least squares estimation: although some rejection rules have been derived in order to eliminate outliers, their performance is frequently very poor. More robust methods are often necessary, and these are primarily the interest of this paper.

2. Robust methods

It is now widely recognised that the use of the least absolute distance criterion or l_1 norm is of value as it tends to give less weight to the effect of gross errors or wild points. In this case

$$\psi(\mathbf{r}) = \sum_{i=1}^{m} |r_i|,$$

the l_1 norm of $\mathbf{r}$. Then the subproblem (1.5) can be expressed as

$$\text{minimize } \|\mathbf{r} + A\,\mathbf{h}\|_1 + \tfrac{1}{2}\mathbf{h}^T B\,\mathbf{h} \tag{2.1}$$

$$\text{subject to } \|\mathbf{h}\|_\infty \le \tau.$$

The choice of the l_∞ norm for the constraint is a natural one, which permits the constraint to be expressed as a set of simple bounds on the components of h. Methods for the nonlinear l_1 problem based on this subproblem, or variants of it, are now well established and can be very successful, with superlinear convergence obtained in very general circumstances. However, there are some disadvantages: (i) such methods are quite sophisticated, for example in the way in which B is chosen, (ii) the solution of the resulting subproblem (2.1) is not straightforward, and needs non-standard software, (iii) good software packages for the whole solution process are not readily available. In addition, there may be reasons to do with the size of the problem, for example, why it is not appropriate to calculate, or work with, an appropriate matrix B.

If B can be set to zero in (2.1), these difficulties are largely removed: (2.1) is just a linear l_1 problem with bound constraints, for which good software is available [2]; the only other matter is the adjustment of τ at each iteration, and there are well established rules for doing this [3]. The price to be paid for using this simpler trust region based approach is that convergence can be slow: it is well known that if $\mathbf{r}$ has n zero components at a stationary point, then second order convergence with B=0 can occur to that point, but otherwise slow linear convergence, with the trust region radius always active, will occur. In addition, unlike the linear problem, there is no particular reason why there should be n zero components of $\mathbf{r}$ at stationary points.

The following question arises, therefore. Can the trust region method (with B=0) form the basis of an algorithm, but can relatively inexpensive and easily programmed modifications be introduced to improve the rate of convergence, when this is slow? Some recent work by Jonasson and Madsen [4] for the corresponding l_∞ problem have suggested that some improvement can be gained by incorporating a correction step, and we consider the analogue of that here for the l_1 problem. Assume for the moment that it is known that at the stationary point $\mathbf{a}^*$ to which the process is converging, $\mathbf{r}(\mathbf{a}^*)$ is zero in indices $i \in I$, where I contains $k<n$ integers. Let a step of the main method be taken which gives a new point $\mathbf{a}+\mathbf{h}$, and assume that a correction step $\mathbf{v}$ is now attempted. In particular, it would seem to be of value to choose $\mathbf{v} \in \mathbb{R}^n$ so that

$$r_i(\mathbf{a}+\mathbf{h}+\mathbf{v}) = 0, \quad i \in I.$$

Because there is still freedom in $\mathbf{v}$, a natural way to remove that is to choose $\mathbf{v}$ as small as possible, for example by solving

$$\text{minimize } \tfrac{1}{2}\mathbf{v}^T\mathbf{v} \text{ subject to } r_i(\mathbf{a}+\mathbf{h}+\mathbf{v}) = 0, \quad i \in I.$$

Of course, the constraints are nonlinear, so an approximation which gives a more tractable problem is given by

$$\text{minimize } \tfrac{1}{2}\mathbf{v}^T\mathbf{v} \text{ subject to } r_i(\mathbf{a}+\mathbf{h})+\mathbf{v}^T\nabla r_i(\mathbf{a}+\mathbf{h}) = 0, \quad i \in I.$$

Finally, in the interests of economy, and assuming $\mathbf{h}$ is small, the fact that the partial derivatives are already available at $\mathbf{a}$ suggests solving

$$\text{minimize } \tfrac{1}{2}\mathbf{v}^T\mathbf{v} \text{ subject to } r_i(\mathbf{a}+\mathbf{h})+\mathbf{v}^T\nabla r_i(\mathbf{a}) = 0, \quad i \in I. \tag{2.2}$$

This problem can be solved relatively cheaply (compared to an iteration of the main method). Various ways of incorporating (2.2) have been tried. Following the idea of Jonasson and Madsen [4], a correction step was attempted when the basic step was regarded as poor, resulting in the new point $\mathbf{a}+\mathbf{h}+\mathbf{v}$ in case of a successful correction. In contrast to this, the use of (2.2) was tried whenever the basic step gave sufficient improvement. The motivation for this is that in this case we assume the guess for the index set I to be a good one and want to make use of this by adding an extra step with the same indices in I. As a variation of the first version the correction was tried *instead* of a poor step, so that $\mathbf{a}+\mathbf{h}$ has to be replaced by $\mathbf{a}$ in (2.2): the new point would then be taken to be $\mathbf{a}+\mathbf{v}$ if the correction was acceptable. The evidence of some numerical tests is that improvement can be obtained in the number of iterations required to attain a certain convergence criterion, although this is not always the case, and the improvement need not be a very big one. The most successful modification appeared to be that which involved the incorporation of a correction step after a good basic step.

The fact that the l_1 norm is non-differentiable is often seen as a disadvantage to its use, and other robust loss functions have been suggested which are smooth. An example is the Huber M-estimator ([5], [6]) where, for example, we can take

$$\psi(\mathbf{r}) = \sum_{i=1}^{m} \rho(r_i/k\sigma), \tag{2.3}$$

where

$$\rho(t) = \begin{cases} \tfrac{1}{2}t^2, & |t| \le 1 \\ |t| - \tfrac{1}{2}, & |t| > 1, \end{cases} \tag{2.4}$$

and k and σ are scaling parameters. Clearly if the product $k\sigma$ is large enough, then (2.3) just gives the l_2 norm, but as that product tends to zero, then the l_1 norm is given in the limit. Thus (2.3) may be seen as a compromise between those two norms. For details of the statistical significance of the scaling parameters, see Huber [5]. Methods for minimising (2.3) based on the Levenberg-Marquardt method have been considered, for example, in [7]. They appear to work well, although the rate of convergence is slow in certain cases.

3. Robust methods for the errors-in-variables problem

Examination of the model equations (1.2) shows that for any **a**, the number r_i is just the vertical distance from (x_i, y_i) to the point $(x_i, f(x_i, \mathbf{a}))$ on the model curve. Thus, the use of the loss function in (1.3) reflects the assumption that only the values of y_i are in error, and that the x_i values are exact. However, this is often not the case, and the independent variable values may also contain errors which cannot be simply ignored: this is normally the case if both variable values come from genuine observations. This situation gives rise to the so-called errors-in-variables problem, and suggests that the correct model equations are

$$y_i + r_i = f(x_i + \varepsilon_i, \mathbf{a}), \quad i=1,..,m. \tag{3.1}$$

In this case (1.3) is replaced by

$$\text{find } \mathbf{a} \in I\!R^n \text{ to minimize } \psi_1(\mathbf{r}) + \psi_2(\varepsilon), \tag{3.2}$$

where ψ_1 and ψ_2 are (possibly different) loss functions. For example if both functions are just the sum of squares, then what is being minimized is the sum of squared orthogonal distances between each data point and the curve described by the model. This is referred to as orthogonal distance regression, and a stable and efficient algorithm for the problem is described by Boggs et al [8] based on a Levenberg-Marquardt trust region approach. Software is available in the form of the Fortran subroutine ODRPACK [9].

In the rest of this paper, we are concerned with (3.2) when the loss functions are the l_1 norm and the Huber M-estimator. In the former case, (3.2) becomes

$$\text{find } \mathbf{a} \in I\!R^n \text{ to minimize } \|\mathbf{g}\|_1, \tag{3.3}$$

where

$$g_i = r_i, \quad i=1,..,m$$

$$g_{i+m} = \varepsilon_i, \quad i=1,..,m.$$

Let G be the matrix of partial derivatives of the vector **g** with repect to the variables

$\eta \in \mathbb{R}^{m+n}$, where

$$\eta^T = (\mathbf{a}^T, \ \varepsilon^T).$$

Then it is readily seen that

$$G = \begin{bmatrix} A & D \\ 0 & I_m \end{bmatrix} \in \mathbb{R}^{2m \times (m+n)}, \tag{3.4}$$

where A is as before, and the (i,j) components of D are given by

$$D_{ij} = \frac{\partial r_i}{\partial \varepsilon_j} = \frac{\partial f_i}{\partial \varepsilon_j}, \quad i,j=1,..,m.$$

Notice that D is a diagonal matrix.

Now consider the application of a Levenberg-Marquardt trust region method. Then the vector $\mathbf{g}$ in (3.3) is replaced by a linear approximation, and the Levenberg-Marquardt step is now obtained by solving a problem having the form

$$\begin{aligned} &\text{minimize } \left\| \begin{bmatrix} \mathbf{r} \\ \varepsilon \end{bmatrix} + G \begin{bmatrix} \delta \mathbf{a} \\ \delta \varepsilon \end{bmatrix} \right\|_1 \\ &\text{subject to } -\tau_1 \le \delta \mathbf{a} \le \tau_1, \\ &\qquad -\tau_2 \le \delta \varepsilon \le \tau_2, \end{aligned} \tag{3.5}$$

where the vectors τ_1 and τ_2 are given. The elements of both τ_1 and τ_2 would normally be set to a common value, but there may be situations where the extra flexibility is desirable. Now by a standard argument, this problem can be converted into a linear programming problem by the introduction of $4m$ additional non-negative variables forming the vectors $\mathbf{u}_1$, $\mathbf{u}_2$, $\mathbf{v}_1$ and $\mathbf{v}_2$, all of which are in $\mathbb{R}^m$. Then, (3.5) is equivalent to the bounded variable linear programming problem

$$\text{mimimize } \mathbf{e}^T(\mathbf{u}_1 + \mathbf{u}_2 + \mathbf{v}_1 + \mathbf{v}_2)$$

subject to

$$\mathbf{u}_1 - \mathbf{v}_1 = \mathbf{r} + A\,\delta\mathbf{a} + D\,\delta\varepsilon,$$

$$\mathbf{u}_2 - \mathbf{v}_2 = \varepsilon + \delta\varepsilon,$$

$$-\tau_1 \le \delta\mathbf{a} \le \tau_1,$$

$$-\tau_2 \le \delta\varepsilon \le \tau_2,$$

$$\mathbf{u}_1, \mathbf{u}_2, \mathbf{v}_1, \mathbf{v}_2 \ge 0.$$

Clearly m of the equality constraints together with $\delta\varepsilon$ can be eliminated directly and so the problem can immediately be simplified to

$$\text{mimimize } \mathbf{e}^T(\mathbf{u}_1 + \mathbf{u}_2 + \mathbf{v}_1 + \mathbf{v}_2)$$

subject to

$$[I\ -I\ -D\ D\ -A]\begin{bmatrix}\mathbf{u}_1\\ \mathbf{v}_1\\ \mathbf{u}_2\\ \mathbf{v}_2\\ \delta\mathbf{a}\end{bmatrix} = \mathbf{r} - D\varepsilon,$$

$$-\tau_1 \le \delta\mathbf{a} \le \tau_1,$$

$$-\tau_2 \le \mathbf{u}_2 - \mathbf{v}_2 - \varepsilon \le \tau_2,$$

$$\mathbf{u}_1, \mathbf{u}_2, \mathbf{v}_1, \mathbf{v}_2 \ge 0.$$

In fact the apparently inconvenient inequality constraints can be eliminated, and replaced by simple bound constraints on the components of $\mathbf{u}_2$ and $\mathbf{v}_2$, leaving a subproblem of a similar size to that which would be required for the usual l_1 problem. For details see [10], where numerical results of the application of the overall method to some problems are given.

Consider now the use of the Huber M-estimator. In this case (3.2) may be written as

$$\text{find } \eta \in I\!R^{m+n} \text{ to minimize } \sum_{i=1}^{2m} \rho(g_i(\eta)/k\sigma),$$

where $\mathbf{g}$, η are as before, and ρ is defined by (2.4). The minimization of this function (*inter alia*) is considered in [11], where a local method based on Newton's method is given. We will show in the rest of the paper how the structure of the problem can again be exploited in the implementation of Levenberg-Marquardt type methods. Let $\mathbf{z}^T = (\delta\mathbf{a}^T, \delta\varepsilon^T)$, let a linearization of $g_i(\eta+\mathbf{z})$ be given by

$$l_i(\mathbf{z}) = g_i(\eta) + \nabla g_i(\eta)\mathbf{z}, \quad i=1,..,2m,$$

and let

$$L(\mathbf{z}) = \sum_{i=1}^{2m} \rho(l_i(\mathbf{z})/k\sigma),$$

where the dependence on η is not shown explicitly. Then the usual trust region step is given by

$$\text{minimize } L(\mathbf{z}) \text{ subject to } \|\mathbf{z}\|_2 \le \tau, \tag{3.6}$$

using the l_2 norm for the trust region radius. Now if equality is assumed to hold in the constraint, we could try to find a zero of the corresponding Lagrangian function for (3.6), and at the same time satisfy the constraint, by a Newton iteration. However, if instead of assuming that τ is known, we assume that the Lagrange multiplier λ is known, then the Newton iteration simplifies to

$$[\nabla^2 L(\mathbf{z}) + 2\lambda I_{m+n}]\delta\mathbf{z} = -[\nabla L(\mathbf{z}) - 2\lambda\mathbf{z}]. \tag{3.7}$$

If necessary, a line search may be used if a good approximation to the solution is not available, so that the process is not dependent on the proximity of the initial approximation. Note that the value of the corresponding trust region radius τ is then just given by $\|\mathbf{z}\|_2$ at the solution.

We will conclude this section by indicating how (3.7) can be efficiently solved, by exploiting the problem structure so that it is essentially equivalent to the solution of an $n\times n$ system rather than an $(m+n)\times(m+n)$ system. It is readily seen that

$$\nabla L(\mathbf{z}) = \sum_{i=1}^{2m} v_i(\mathbf{z})\nabla l_i(\mathbf{z})/k\sigma,$$

where

$$v_i(\mathbf{z}) = \rho'(l_i(\mathbf{z})/k\sigma), \quad i=1,..,2m,$$

and the dash just refers to differentiation of ρ with respect to its argument. Thus $v_i(\mathbf{z})=\pm1$, if $|l_i(\mathbf{z})|>k\sigma$, and $l_i(\mathbf{z})/k\sigma$ otherwise. Let G be the $2m\times(m+n)$ matrix whose rows are $\nabla g_i(\eta)$, so that G is given as before by (3.4). It follows that

$$\nabla L(\mathbf{z}) = G^T \mathbf{v}(\mathbf{z})/k\sigma,$$

$$\nabla^2 L(\mathbf{z}) = G^T W(\mathbf{z}) G/k^2\sigma^2,$$

where $W = \mathrm{diag}\{\rho''(l_i(\mathbf{z})/k\sigma),\ i=1,..,2m\}$. Thus $w_i=0$ if $|l_i(\mathbf{z})|>k\sigma$, and $w_i=1$ otherwise. Let the leading $m\times m$ submatrix of W be W_1, and the bottom $m\times m$ right hand submatrix be W_2. Then using the form of G given by (3.4), it follows that (3.7) can be rewritten in partitioned form, say as

$$\begin{bmatrix} S & T^T \\ T & V \end{bmatrix}\begin{bmatrix} \delta\mathbf{z}_1 \\ \delta\mathbf{z}_2 \end{bmatrix} = \begin{bmatrix} \mathbf{c} \\ \mathbf{d} \end{bmatrix}. \tag{3.8}$$

where, in particular, $\delta\mathbf{z}^T = (\delta\mathbf{z}_1^T, \delta\mathbf{z}_2)^T$, and

$$S = A^T W_1 A + 2\lambda k^2\sigma^2 I_n,$$

$$T = DW_1 A,$$

and

$$V = DW_1 D + W_2 + 2\lambda k^2\sigma^2 I_m.$$

Note that V is an $m\times m$ diagonal matrix. Further, it is nonsingular (because λ can be kept positive), and so $\delta\mathbf{z}_2$ can be eliminated from (3.8) using

$$\delta\mathbf{z}_2 = V^{-1}(\mathbf{d} - T\delta\mathbf{z}_1), \tag{3.9}$$

and $\delta\mathbf{z}_1$ can be obtained by solving the $n\times n$ system

$$(S-T^TV^{-1}T)\delta\mathbf{z}_1 = \mathbf{c}-T^TV^{-1}\mathbf{d}. \tag{3.10}$$

It follows that the Newton direction can be calculated for the errors-in-variables problem in essentially the same amount of computation as for the usual problem. Finally, we summarize the algorithm.

1. Choose initial approximations η (or $\mathbf{a}$, ε), and an initial value of λ.
2. (Inner iteration for $\mathbf{z}$: Newton's method with line search.)

   Choose an initial approximation for $\mathbf{z}$ and solve for $\delta\mathbf{z}$ satisfying (3.7), but using (3.10) and (3.9). Update $\mathbf{z}$ to $\mathbf{z}+\alpha\delta\mathbf{z}$, where the scalar α is set to 1 if possible, but may need to have a value less than that, and repeat until convergence.
3. If $\eta+\mathbf{z}$ gives an acceptable decrease in the Huber M-estimator function, replace η by $\eta+\mathbf{z}$, and update λ. Otherwise, increase λ (this corresponds to a reduction in the trust region radius τ of (3.6)).
4. Return to Step 2 unless a stopping criterion is met.

Various other aspects of this procedure, and numerical results for its application to data fitting problems, are part of the PhD work presently being undertaken by the first author.

4. Implicit models

Writing $\mathbf{x} = (x, y)^T$, then (1.1) can be written in the form

$$F(\mathbf{x}, \mathbf{a}) = 0. \tag{4.1}$$

Normally, there would be no advantage in doing this. However, for some problems, (4.1) is the natural representation of the relationship between the variables, and indeed it may not be possible to identify one of the variables as a dependent variable, as in (1.1). An example is an ellipse described by the relation

$$x_1^2/a_1^2 + x_2^2/a_2^2 = 1.$$

Such models are known as *implicit models*, and it is common for all the variables to contain errors. Let $\mathbf{x} \in \mathbb{R}^k$. Then introducing a vector ε_i of errors into observed values of $\mathbf{x}_i$, $i=1,..,m$, leads to the problem

$$\text{minimize } \psi(\varepsilon)$$

$$\text{subject to } F(\mathbf{x}_i + \varepsilon_i, \mathbf{a}) = 0, \quad i=1,..,m,$$

where ε denotes the vector in $\mathbb{R}^{mk}$ given by concatenating the m vectors $\varepsilon_i \in \mathbb{R}^k$, and ψ is a suitable loss function. This is an equality constrained optimization problem in $n+mk$ variables. Some attempts to solve least squares problems of this kind using a Gauss-Newton approach are given in [12]. A modification of Levenberg-Marquardt type has been presented in [13]. It would appear that little attention has been given to the use of more robust loss functions for such problems, although locally convergent methods are discussed in [11]. It is intended to consider globally convergent methods, which make full use of the structure of the problem, in future work.

References

1. G. A. Watson, in The State of the Art in Numerical Analysis, A. Iserles and M. J. D. Powell (eds), Clarendon Press, Oxford (1987) pp.139-164.
2. I. Barrodale and F. D. K. Roberts, SIAMJ Num Anal, 15, pp.603-611 (1978).
3. R. Fletcher, Practical Methods of Optimization, John Wiley, New York (1987).
4. K. Jonasson and K. Madsen, Technical University of Denmark Report, NI-92-06 (1992).
5. P. Huber, Anals of Statistics, 1, pp.799-821 (1973).
6. P. Huber, Robust Statistics, John Wiley, New York (1981).
7. G. Li and K. Madsen, in: Numerical Analysis 1987, D. F. Griffiths and G. A. Watson (eds), Pitman Research Notes in Mathematics 170, Longman, Essex (1988) pp.176-191.
8. P. T. Boggs, R. H. Byrd and R. B. Schnabel, SIAM J Sci Stat Comp, 8, pp.1052-1078 (1987).
9. P. T. Boggs, R. H. Byrd, J. R. Donaldson and R. B. Schnabel, ACM Trans Math Software, 15, pp.348-364 (1989).
10. G. A. Watson and K. F. C. Yiu, BIT, 31, pp.697-710 (1991).
11. W. H. Jeffreys, Biometrika, 77, pp.597-607 (1990).
12. G. A. Watson , in Multivariate Approximation Theory III, W. Schempp and K. Zeller (eds), ISNM 75, Birkhauser Verlag, Basel (1985) pp.388-400.
13. D. Zwick, paper presented at the NATO Advanced Research Workshop, Algorithms for Approximation, Oxford (1992).

2nd Int. Coll. on Numerical Analysis, pp. 103-112
D. Bainov and V. Covachev (Eds)

INTERPOLATION OF FUNCTION SPACES AND THE CONVERGENCE RATE ESTIMATES FOR THE FINITE DIFFERENCE SCHEMES

Boško S. Jovanović

University of Belgrade, Faculty of Sciences,
Studentski trg 16, POB 550, 11000 Belgrade, Yugoslavia

Abstract: In this paper we show how the theory of interpolation of function spaces can be used to obtain the convergence rate estimates for the finite difference schemes. In particular, in the case of initial-boundary value problems of hyperbolic type, this method yields new, nonstandard, fractional order estimates.

MS Classifications: 65N15, 46B70

Key Words: Boundary-Value Problems, Finite Differences, Interpolation of function spaces, Sobolev Spaces, Convergence Rate Estimates

1. Introduction

For a wide class of finite difference schemes for elliptic boundary value problems, the estimates of the convergence rates cosistent with the smoothness of data [18, 12–14, 5, 10], are of major interest, i.e.

$$\|u-v\|_{W_{2,h}^k} \le C\,h^{s-k}\,\|u\|_{W_2^s(\Omega)}\,, \qquad s>k\,. \tag{1}$$

Here u denotes the solution of the original boundary value problem, v denotes the solution of the corresponding finite difference scheme, h is the discretisation parameter, $W_2^s(\Omega)$ denotes the Sobolev space, $W_{2,h}^k$ denotes the discrete Sobolev space, and C is a positive generic constant, independent of h and u.

Analogous estimates hold in the parabolic case [6, 7]:

$$\|u-v\|_{W_{2,h\,\tau}^{k,k/2}} \le C\,h^{s-k}\,\|u\|_{W_2^{s,s/2}(Q)}\,, \qquad s>k\,. \tag{2}$$

On the contrary, in the hyperbolic case, we only have weak estimates, not compatible with the smoothness of data [9, 8]:

$$\|u-v\|_{C_\tau(W_{2,h}^k)} \le C\,h^{s-k-1}\,\|u\|_{W_2^s(Q)}\,, \qquad s>k+1\,. \tag{3}$$

Estimates (1–3) are usually obtained using the Bramble–Hilbert lemma [3, 4].

2. Some basic results of the theory of interpolation of function spaces

Let A_1 and A_2 be two Banach spaces, linearly and continuously embedded in a topological linear space $\mathcal{A}$. Two such spaces are called *interpolation pair* $\{A_1, A_2\}$. Consider also the space $A_1 \cap A_2$, with the norm

$$\|a\|_{A_1 \cap A_2} = \max\{\|a\|_{A_1}, \|a\|_{A_2}\}$$

and the space $A_1 + A_2 = \{a \in \mathcal{A} : a = a_1 + a_2, \ a_j \in A_j, \ j = 1, 2\}$, with the norm

$$\|a\|_{A_1+A_2} = \inf_{\substack{a=a_1+a_2 \\ a_j \in A_j}} \{\|a_1\|_{A_1} + \|a_2\|_{A_2}\} .$$

Obviously, $A_1 \cap A_2 \subset A_j \subset A_1 + A_2$, $j = 1, 2$.

Let us introduce the category $\mathcal{C}_1$, where the objects $A, B, C, \ldots$ are Banach spaces, and morphisms — bounded linear operators $L \in \mathcal{L}(A, B)$. Let, also, $\mathcal{C}_2$ be a category where the objects are interpolation pairs $\{A_1, A_2\}, \{B_1, B_2\}, \ldots$ while morphisms are $L \in \mathcal{L}(\{A_1, A_2\}, \{B_1, B_2\})$. Here $\mathcal{L}(\{A_1, A_2\}, \{B_1, B_2\})$ denotes the set of bounded linear operators from $A_1 + A_2$ into $B_1 + B_2$, whose restrictions on A_j belong to the set $\mathcal{L}(A_j, B_j)$, $j = 1, 2$.

Functor $\mathcal{F} : \mathcal{C}_2 \to \mathcal{C}_1$ is called *interpolation functor* if

$$A_1 \cap A_2 \subset \mathcal{F}(\{A_1, A_2\}) \subset A_1 + A_2$$

for every interpolation pair $\{A_1, A_2\}$, while for every morphism $L \in \mathcal{L}(\{A_1, A_2\}, \{B_1, B_2\})$, $\mathcal{F}(L)$ is the restriction of the operator L on $\mathcal{F}(\{A_1, A_2\})$.

The corresponding Banach space $A = \mathcal{F}(\{A_1, A_2\})$ is called *interpolation space.*

Note that $A_1 \cap A_2$ and $A_1 + A_2$ are interpolation spaces.

If the inequality

$$\|L\|_{\mathcal{F}(\{A_1, A_2\}) \to \mathcal{F}(\{B_1, B_2\})} \le C \, \|L\|^{1-\theta}_{A_1 \to B_1} \, \|L\|^{\theta}_{A_2 \to B_2} ,$$
$$0 < \theta < 1 , \qquad C = \text{const} \ge 1$$

is satisfied, the interpolation functor $\mathcal{F}$ is called to be of the *type* θ. (Here $\|L\|_{A_j \to B_j}$ denote the usual operator norm of $L : A_j \to B_j$).

One of the most often used interpolation methods is so called K–method [15, 2]. Let $\{A_1, A_2\}$ be an interpolation pair. Define the funcion

$$K(t, a, A_1, A_2) = \inf_{\substack{a \in A_1+A_2 \\ a=a_1+a_2 \\ a_j \in A_j}} \{\|a_1\|_{A_1} + t \, \|a_2\|_{A_2}\} .$$

It is obvious, that for a fixed $t \in (0, \infty)$, $K(t, a, A_1, A_2)$ is a norm in $A_1 + A_2$, equivalent to the standard norm $\|a\|_{A_1+A_2}$. For $0 < \theta < 1$ and $1 \le q < \infty$ let us define the space $(A_1, A_2)_{\theta, q}$ as the set of all elements $a \in A_1 + A_2$ with the finite norm

$$\|a\|_{(A_1,\,A_2)_{\theta,\,q}} = \left\{ \int_0^\infty \left[t^{-\theta}\, K(t,\, a,\, A_1,\, A_2) \right]^q \frac{dt}{t} \right\}^{1/q}.$$

Such defined space $(A_1,\, A_2)_{\theta,\,q}$ is an interpolation space. The following relations hold

$$\begin{aligned} &(A_1,\, A_2)_{\theta,\,q} = (A_2,\, A_1)_{1-\theta,\,q}\,, \\ &(A,\, A)_{\theta,\,q} = A\,, \\ &\|a\|_{(A_1,\,A_2)_{\theta,\,q}} \le C_{\theta,\,q}\, \|a\|_{A_1}^{1-\theta}\, \|a\|_{A_2}^{\theta}\,, \qquad \forall\, a \in A_1 \cap A_2\,. \end{aligned}$$

The corresponding interpolation functor $\mathcal{F}(\{A_1,\, A_2\}) = (A_1,\, A_2)_{\theta,\,q}$ is of the type θ, i.e.

$$\|L\|_{(A_1,\,A_2)_{\theta,\,q}\to(B_1,\,B_2)_{\theta,\,q}} \le \|L\|_{A_1\to B_1}^{1-\theta}\, \|L\|_{A_2\to B_2}^{\theta}\,. \tag{4}$$

As an example of interpolation function spaces, let us consider the Sobolev spaces W_p^s [1, 17]. For $1 < p < \infty$ and $s = 0,\, 1,\, 2,\, \ldots$ they are defined by means of the Lebesgue spaces L_p

$$W_p^s(\mathbb{R}^n) = \Big\{ f \in L_p(\mathbb{R}^n) \,:\, \|f\|_{W_p^s(\mathbb{R}^n)} = \Big(\sum_{|\alpha|\le s} \|D^\alpha f\|_{L_p(\mathbb{R}^n)}^p \Big)^{1/p} < \infty \Big\}.$$

Here, as usual, $\alpha = (\alpha_1,\, \ldots\,,\, \alpha_n)$ is a multi-index, $|\alpha| = \alpha_1 + \cdots + \alpha_n$ and $D^\alpha = (\partial/\partial x_1)^{\alpha_1} \cdots (\partial/\partial x_n)^{\alpha_n}$ — partial derivative. For noninteger positive s one set

$$W_p^s(\mathbb{R}^n) = B_{pp}^s(\mathbb{R}^n)\,,$$

where B_{pq}^s is the Besov space [17].

For $0 \le s_1,\, s_2 < \infty\,,\ s_1 \ne s_2\,,\ 0 < \theta < 1\,,\ 1 \le q < \infty$ we have [17]

$$\big(W_p^{s_1}(\mathbb{R}^n),\, W_p^{s_2}(\mathbb{R}^n)\big)_{\theta,\,q} = B_{pq}^s(\mathbb{R}^n)\,, \qquad s = (1-\theta)\, s_1 + \theta\, s_2\,.$$

In such a way, for $q = p$ and noninteger $s = (1-\theta)\, s_1 + \theta\, s_2$, we obtain

$$\big(W_p^{s_1}(\mathbb{R}^n),\, W_p^{s_2}(\mathbb{R}^n)\big)_{\theta,\,p} = W_p^s(\mathbb{R}^n)\,, \qquad s = (1-\theta)\, s_1 + \theta\, s_2\,.$$

For $p = 2$, this relation holds without restrictions, i.e.

$$\big(W_2^{s_1}(\mathbb{R}^n),\, W_2^{s_2}(\mathbb{R}^n)\big)_{\theta,\,2} = W_2^{(1-\theta)\, s_1 + \theta\, s_2}(\mathbb{R}^n)\,.$$

Hence, $W_2^s(\mathbb{R}^n)$ are interpolation spaces. The same result holds for the Sobolev spaces in a domain Ω with sufficiently smooth boundary.

3. Application of the interpolation theory to the determination of error bounds for difference schemes: elliptic case

Interpolation theory can be used to determine the error bounds for the finite difference schemes. Let us consider, for example, the Dirichlet boundary value problem for the Poisson equation

$$(5)\qquad \begin{aligned} -\Delta u &= f \quad \text{in} \quad \Omega = (0,1)^n\,, \\ u &= 0 \quad \text{on} \quad \Gamma = \partial\Omega\,. \end{aligned}$$

In the domain $\overline{\Omega} = \Omega \cup \Gamma$ let us define an uniform mesh $\overline{\omega}_h$ with the step size h. Denote $\omega_h = \overline{\omega}_h \cap \Omega$, $\gamma_h = \overline{\omega}_h \cap \Gamma$. Let v_{x_j} and $v_{\bar{x}_j}$ denote the upward and backward finite differences in the variable x_j. We also define the discrete analogues $\|\cdot\|_{L_{2,h}}$, $\|\cdot\|_{W^k_{2,h}}$, $|\cdot|_{W^k_{2,h}}$ of the Lebesgue and Sobolev norms and seminorms.

The boundary value problem (5) we approximate with the "cross" difference scheme with the averaged right hand side

$$(6)\qquad \begin{aligned} -\Delta_h v \equiv -\sum_{j=1}^{n} v_{x_j\bar{x}_j} &= S_1 \cdots S_n f \quad \text{in} \quad \omega_h\,, \\ v &= 0 \quad \text{on} \quad \gamma_h\,. \end{aligned}$$

Here S_j is the Steklov averaging operator in x_j

$$S_j f(x_1, \ldots, x_j, \ldots, x_n) = \int_{x_j-h/2}^{x_j+h/2} f(x_1, \ldots, \xi_j, \ldots, x_n)\, d\xi_j\,.$$

The error $z = u - v$, for $n = 2, 3$, satisfies estimate [5]

$$(7)\qquad \|z\|_{W^2_{2,h}} \le C\, h^2\, \|u\|_{W^4_2(\Omega)}\,.$$

The inequality

$$(8)\qquad \|z\|_{W^2_{2,h}} \le C\, \|u\|_{W^2_2(\Omega)}$$

holds too.

Let us consider the mapping $z = Lu$. Operator L is, obviously, linear. From (7) it follows that L is a bounded operator from $W^4_2(\Omega)$ to $W^2_{2,h}$, and

$$\|L\|_{W^4_2(\Omega)\to W^2_{2,h}} \le C\, h^2\,.$$

Analogously, from (8) follows that L is a bounded operator from $W^2_2(\Omega)$ to $W^2_{2,h}$, and

$$\|L\|_{W^2_2(\Omega)\to W^2_{2,h}} \le C\,.$$

By interpolation, from inequality (4), we directly conclude that L is a bounded linear operator from $\left(W^2_2(\Omega),\, W^4_2(\Omega)\right)_{\theta,2} \equiv W^{2+2\theta}_2(\Omega)$ into $W^2_{2,h}$ and

$$\|L\|_{W^{2+2\theta}_2(\Omega)\to W^2_{2,h}} \le C\, h^{2\theta}\,.$$

Hence

$$\|z\|_{W_{2,h}^2} = \|Lu\|_{W_{2,h}^2} \le \|L\|_{W_2^{2+2\theta}(\Omega)\to W_{2,h}^2} \|u\|_{W_2^{2+2\theta}(\Omega)}$$
$$\le C\,h^{2\theta}\,\|u\|_{W_2^{2+2\theta}(\Omega)}\,, \qquad 0<\theta<1\,.$$

Setting $2+2\theta=s$, we finally obtain

$$\|z\|_{W_{2,h}^2} \le C\,h^{s-2}\,\|u\|_{W_2^s(\Omega)}\,, \qquad 2\le s\le 4\,.$$

Analogously, from the integer order estimates of the type (1), we can obtain the fractional order estimates. Similar results hold true in the parabolic case.

4. Application of the interpolation theory to the determination of error bounds for difference schemes: hyperbolic case

Let us consider, as a model problem, the first initial–boundary value problem for the wave equation in the cylinder $Q=\Omega\times(0,T)=(0,1)^n\times(0,T)$

$$\begin{aligned} &\partial^2 u/\partial t^2 = \Delta u\,, && (x,t)\in Q\,,\\ &u(x,t)=0\,, && (x,t)\in\Gamma\times[0,T]\,,\\ &u(x,0)=u_0(x)\,, && \partial u(x,0)/\partial t=0\,, \qquad x\in\Omega\,. \end{aligned} \tag{9}$$

In the sequel, we assume that $u_0\in W_2^s(\Omega,\ s\ge 1$, and can be oddly extended outside Ω, preserving the class. Let $C(L_2)$, $C(W_2^s)$ and $L_q(W_2^s)$ be the spaces of functions defined on $[0,T]$ with the values in $L_2(\Omega)$, or $W_2^s(\Omega)$, with the norms

$$\|u\|_{C(L_2)} = \max_{t\in[0,T]} \|u(t)\|_{L_2(\Omega)}\,, \qquad \|u\|_{C(W_2^s)} = \max_{t\in[0,T]} \|u(t)\|_{W_2^s(\Omega)}$$
$$\text{and} \qquad \|u\|_{L_q(W_2^s)} = \|\,\|u(t)\|_{W_2^s(\Omega)}\|_{L_q(0,T)}\,.$$

The solution of the initial–boundary value problem (9) satisfies the a priori estimate [11]

$$\|u\|_{C(W_2^1)} \le C\,|u_0|_{W_2^1(\Omega)}\,, \tag{10}$$

from which, differentiating equation (9), follows

$$\left\|D^\alpha u\right\|_{C(L_2)} \le C\,|u_0|_{W_2^k(\Omega)}\,, \qquad |\alpha|=k\,, \qquad 1\le k\le[s]\,. \tag{11}$$

Besides the mesh $\overline{\omega}_h$ introduced earlier, let $\overline{\omega}_\tau$ be an uniform mesh with the step size τ on the interval $[-\tau/2,T]$. Denote $\omega_\tau=\overline{\omega}_\tau\cap(0,T)$ and $\omega_\tau^-=\omega_\tau\cup\{-\tau/2\}$. In the sequel we shall use the following notation

$$v=v(t)\,, \quad \hat v = v(t+\tau)\,, \quad \check v = v(t-\tau)\,, \quad v^j = v((j-1/2)\tau)\,,$$

$$\overline{v}=(v+\hat v)/2\,, \quad v_t=(\hat v - v)/\tau\,, \quad v_{\bar t}=(v-\check v)/\tau\,.$$

For the funcions defined on the mesh $\overline{\omega}_h \times \overline{\omega}_\tau$ we define the norms

$$\|v\|_{C_\tau(L_{2,h})} = \max_{t\in\omega_\tau^-} \|v(\cdot,t)\|_{L_{2,h}}\,, \qquad \|v\|_{C_\tau(W^k_{2,h})} = \max_{t\in\omega_\tau^-} \|v(\cdot,t)\|_{W^k_{2,h}}$$

$$\text{and} \qquad \|v\|_{L_{q,\tau}(L_{2,h})} = \Big\{\tau \sum_{t\in\omega_\tau} \|v(\cdot,t)\|^q_{L_{2,h}}\Big\}^{1/q}\,.$$

The initial–boundary value problem (9) we approximate with the following weighted finite difference scheme [16]

$$\begin{aligned}
&v_{t\bar t} = \Delta_h[\sigma\,\hat v + (1-2\sigma)\,v + \sigma\,\check v]\,, \qquad x\in\omega_h\,, \qquad t\in\omega_\tau\,,\\
(12)\qquad &v(x,\,t) = 0\,, \qquad x\in\gamma_h\,, \qquad t\in\overline{\omega}_\tau\,,\\
&v^0 = v^1 = S_1 \cdots S_n u_0(x)\,, \qquad x\in\omega_h\,.
\end{aligned}$$

For $\sigma \geq 1/4$, the difference scheme (12) is absolutely stable and its solution satisfies the relation

$$\|\overline{v}\|_{C_\tau(W^1_{2,\,h})} \leq C\,|v^0|_{W^1_{2,\,h}}\,.$$

The same inequality also holds for $\sigma < 1/4$, if

$$\tau \leq h\sqrt{\frac{1-c_0}{n\,(1-4\,\sigma)}}\,, \qquad c_0 = \text{const} \in (0,1) \qquad \text{(conditional stability).}$$

From initial conditions, using the Newton–Leibnitz formula and Cauchy–Schwartz inequality, we easily obtain

$$|v^0|_{W^1_{2,\,h}} \leq C\,|u_0|_{W^1_2(\Omega)}\,.$$

Hence

$$(13) \qquad \|\overline{v}\|_{C_\tau(W^1_{2,\,h})} \leq C\,\|u_0\|_{W^1_2(\Omega)}\,.$$

Let u be the solution of the initial–boundary value problem (9) and v — the solution of the finite difference scheme (12). Let us define the error as $z = S_1 \cdots S_n u - v$. Such defined error satisfies the following conditions

$$\begin{aligned}
&z_{t\bar t} = \Delta_h[\sigma\,\hat z + (1-2\sigma)\,z + \sigma\check z] + \psi\,, \qquad x\in\omega_h\,, \qquad t\in\omega_\tau\,,\\
&z(x,\,t) = 0\,, \qquad x\in\gamma_h\,, \qquad t\in\overline{\omega}_\tau\,,\\
&z^0 = z^1 = u(x,\,\tau/2) - u_0(x)\,, \qquad x\in\omega_h\,,
\end{aligned}$$

where

$$\psi = S_1\cdots S_n\big\{u_{t\bar t} - \Delta_h[\sigma\,\hat u + (1-2\sigma)\,u + \sigma\,\check u]\big\}\,.$$

The following a priory estimate is satisfied

$$(14) \qquad \|\overline{z}\|_{C_\tau(W^1_{2,\,h})} \leq C\,\big(|z^0|_{W^1_{2,\,h}} + \|\psi\|_{L_{1,\,\tau}(L_{2,\,h})}\big)\,.$$

Using the Newton–Leibnitz formula and the relations

$$u_{x_j\bar x_j} = S_j^2\frac{\partial^2 u}{\partial x_j^2} \qquad \text{and} \qquad u_{t\bar t} = S_t^2\frac{\partial^2 u}{\partial t^2}\,,$$

where S_t is the Steklov averaging operator in t, $z^0_{x_j}$ an ψ can be represented in the following way

$$z^0_{x_j} = S_1 \cdots S_n \Big\{ \frac{1}{h} \int_{x_j}^{x_j+h} \int_0^{\tau/2} \int_0^t \frac{\partial^3 u(x_1, \ldots, \xi_j, \ldots, x_n, \eta)}{\partial t^2\, \partial x_j}\, d\eta\, dt\, d\xi_j \Big\} ,$$

$$\psi = \sum_{j=1}^n S_1 \cdots S_n\, S_t^2 \Big(\frac{\partial^2 u}{\partial x_j^2} - S_j^2 \frac{\partial^2 u}{\partial x_j^2} \Big) - \sum_{j=1}^n S_1 \cdots S_n\, S_j^2 \Big(\frac{\partial^2 u}{\partial x_j^2} - S_t^2 \frac{\partial^2 u}{\partial x_j^2} \Big)$$
$$- \sigma \tau^2 \sum_{j=1}^n S_1 \cdots S_n\, S_j^2\, S_t^2\, \frac{\partial^4 u}{\partial t^2\, \partial x_j^2} .$$

From here, using the identity

$$\varphi - S_j^2 \varphi = \frac{1}{h} \int_{x_j-h}^{x_j+h} \int_{\xi_j}^{x_j} (\xi_j - \eta_j) \Big(1 - \frac{|\xi_j - x_j|}{h} \Big) \frac{\partial^2 \varphi(x_1, \ldots, \eta_j, \ldots, x_n, t)}{\partial x_j^2}\, d\eta_j\, d\xi_j$$

and the Cauchy–Schwartz inequality, we obtain

$$|z^0|_{W^1_{2,h}} \le C \tau^2 \Big| \frac{\partial^2 u}{\partial t^2} \Big|_{C(W_2^1)} \tag{15}$$

and

$$\|\psi\|_{L_{1,\tau}(L_{2,h})} \le C (h^2 + \tau^2) \Big(\Big| \frac{\partial^2 u}{\partial t^2} \Big|_{C(W_2^2)} + |u|_{C(W_2^4)} \Big) . \tag{16}$$

Finally, from (14–16) and (11) we obtain the estimate

$$\|\bar z\|_{C_\tau(W^1_{2,h})} \le C (h + \tau)^2\, \|u_0\|_{W_2^4(\Omega)} . \tag{17}$$

On the other hand,

$$\|\bar z\|_{C_\tau(W^1_{2,h})} \le \|\overline{S_1 \cdots S_n u}\|_{C_\tau(W^1_{2,h})} + \|\bar v\|_{C_\tau(W^1_{2,h})} \le \|u\|_{C(W_2^1)} + \|\bar v\|_{C_\tau(W^1_{2,h})} ,$$

and, using inequalities (10) and (13), from here follows

$$\|\bar z\|_{C_\tau(W^1_{2,h})} \le C\, \|u_0\|_{W_2^1(\Omega)} . \tag{18}$$

From (17) and (18) by interpolation we obtain

$$\|\bar z\|_{C_\tau(W^1_{2,h})} \le C (h + \tau)^{\frac{2}{3}(s-1)}\, \|u_0\|_{W_2^s(\Omega)} , \qquad 1 \le s \le 4 . \tag{19}$$

Similar estimate is obtained in [19] for a projection–difference scheme.

Notice, that for $s = 4$ the estimate (19) reduces to the standard estimate (3) (for $k = 1$), while for $s < 4$ estimate (19) is better. In particular, as opposed to (3), the estimate (19) guarantees the convergence even for $1 < s \le 2$.

Let us consider the fourth order finite difference scheme for the initial–boundary value problem (9)

$$
(20)\qquad
\begin{aligned}
& v_{\bar{t}t} + \frac{h^2 - \tau^2}{12}\,\Delta_h v_{\bar{t}t} = \Delta_h v + \frac{h^2}{6}\sum_{i<j} v_{x_i\bar{x}_i x_j\bar{x}_j}\,, \qquad x \in \omega_h\,, \qquad t \in \omega_\tau\,,\\
& v(x,\, t) = 0\,, \qquad x \in \gamma_h\,, \qquad t \in \overline{\omega}_\tau\,,\\
& v^0 = v^1 = S_1 \cdots S_n u_0 + \frac{\tau^2}{8}\,\Delta_h S_1 \cdots S_n u_0\,, \qquad x \in \omega_h\,.
\end{aligned}
$$

For $n = 2,\, 3$ this scheme is stable for sufficiently small τ ($\tau \le C\,h$). Analogously as in the previous case, we obtain the estimates (18) and

$$\|\overline{z}\|_{C_\tau(W^1_{2,h})} \le C\,h^4\,\|u_0\|_{W^6_2(\Omega)}\,.$$

By interpolation follows

$$\|\overline{z}\|_{C_\tau(W^1_{2,h})} \le C\,h^{\frac{4}{5}(s-1)}\,\|u_0\|_{W^s_2(\Omega)}\,, \qquad 1 \le s \le 6\,.$$

Similar estimates can be obtained in the other discrete norms, for example

$$(21)\qquad \|\overline{z}\|_{C_\tau(L_2)} \le C\,(h+\tau)^{\frac{2}{3}s}\,\|u_0\|_{W^s_2(\Omega)}\,, \qquad 0 \le s \le 3\,,$$

$$\|\overline{z}\|_{C_\tau(W^2_{2,h})} \le C\,(h+\tau)^{\frac{2}{3}(s-2)}\,\|u_0\|_{W^s_2(\Omega)}\,, \qquad 2 \le s \le 5\,,$$

for the second order scheme (12), and

$$\|\overline{z}\|_{C_\tau(L_2)} \le C\,h^{\frac{4}{5}s}\,\|u_0\|_{W^s_2(\Omega)}\,, \qquad 0 \le s \le 5\,,$$

$$\|\overline{z}\|_{C_\tau(W^2_{2,h})} \le C\,h^{\frac{4}{5}(s-2)}\,\|u_0\|_{W^s_2(\Omega)}\,, \qquad 2 \le s \le 7\,,$$

for the fourth order scheme (20).

Presented results can be also transferred to the economic finite difference schemes, obtained by factorisation of the schemes (12) and (20). The scheme (12) is replaced by

$$(I + \sigma\,\tau^2\,\Lambda_1)\cdots(I + \sigma\,\tau^2\,\Lambda_n)\,v_{\bar{t}t} = \Delta_h v\,,$$

and (20) by

$$(I + \frac{\tau^2 - h^2}{12}\,\Lambda_1)\cdots(I + \frac{\tau^2 - h^2}{12}\,\Lambda_n)\,v_{\bar{t}t} = \Delta_h v + \frac{h^2}{6}\sum_{i<j} v_{x_i\bar{x}_i x_j\bar{x}_j}\,, \qquad n = 2,\, 3\,.$$

Here I denote the identity operator, $Iv = v$, and $\Lambda_j v = -v_{x_j\bar{x}_j}$.

Numerical examples

The initial–boundary value problem (9) for $n = 1$ (the equation of vibrating string) was considered in the domain $Q = (0,1) \times (0,\, 1/2)$. We used the finite difference scheme (12) with $\sigma = 1/4$. The initial condition was taken in the form

$$u_0(x) = u_{0l}(x) = \begin{cases} 0\,, & 0 \le x < 1/2\\ \varphi_l(x)\,, & 1/2 \le x \le 1 \end{cases}$$

where

$$\varphi_l(x) = (x - 1/2)^l \, (1 - x) \qquad \text{for} \qquad l = 0,\, 1/2,\, 1,\, 3/2$$

and

$$\varphi_l(x) = (x - 1/2)^l \, (1 - x)^3 \qquad \text{for} \qquad l = 2,\, 5/2\,.$$

Defined like that, $u_0 \in W_2^s(0,\, 1)$, where $s = l + 1/2 - \varepsilon$, with arbitrary small positive value ε.

The order of convergence was determined numerically, according to the estimate (21), using the Runge's rule. The computations was performed on a PC 386. The results are represented in the following table.

Mesh	s					
	0.5	1.0	1.5	2.0	2.5	3.0
16 × 16	0.578	0.464	0.867	0.889	1.277	1.147
32 × 32	0.573	0.754	1.161	1.525	1.226	1.569
64 × 64	0.349	0.749	1.186	1.607	1.667	1.972
128 × 128	0.374	0.749	1.164	1.553	1.822	1.907
256 × 256	0.357	0.720	1.107	1.502	1.781	1.887
$\frac{2}{3}\, s$	0.333	0.667	1.000	1.333	1.667	2.000

Acknowledgements

This research was supported by Science Fund of Serbia, grant number 0401F.

References

[1] R.A. Adams, *Sobolev Spaces*, Academic Press, New York (1975).

[2] J. Bergh, J. Löfström, *Interpolation Spaces*, Springer, Berlin (1976).

[3] J.H. Bramble and S.R. Hilbert, *Numer. Math.* 16, 362–369 (1971).

[4] T. Dupont and R. Scott, *Math. Comput.* 34, 441–463 (1980).

[5] B.S. Jovanović, *in: Variational-Difference Methods in Mathematical Physics, N.S. Bakhvalov, Yu.A. Kuznetsov (Eds.)*, OVM AN SSSR, Moskva (1984), 120–129 (Russian).

[6] B.S. Jovanović, *Numer. Math.* 54, 395–404 (1989).

[7] B.S. Jovanović, *Z. Angew. Math. Mech.* 71, 647–650 (1991).

[8] B.S. Jovanović, *Z. Angew. Math. Mech.* 72, 493–496 (1992).

[9] B.S. Jovanović, L.D. Ivanović and E.E. Süli, *IMA J. Numer. Anal.* 7, 39–45 (1987).

[10] B.S. Jovanović, L.D. Ivanović and E.E. Süli, *IMA J. Numer. Anal.* 7, 301–305 (1987).

[11] O.A. Ladyzhenskaya, *Boudary Value Problems of Mathematical Physics*, Nauka, Moscow (1973) (Russian).

[12] R.D. Lazarov, *Differentsial'nye Uravneniya* 17, 1287–1294 (1981).

[13] R.D. Lazarov, V.L. Makarov and A.A. Samarski, *Mat. Sbornik* 117, 469–480 (1982).

[14] R.D. Lazarov, V.L. Makarov and W. Weinelt, *Numer. Math.* 44, 223–232 (1984).

[15] J. Peetre, *A Theory of Interpolation of Normed Spaces*, Lecture Notes, Brasilia (1963) [Notas de Matematica 39 (1968)].

[16] A.A. Samarski, *Theory of Difference Schemes*, Nauka, Moscow (1983) (Russian).

[17] H. Triebel, *Interpolation Theory, Function Spaces, Differential Operators*, Deutscher Verlag der Wissenschaften, Berlin (1978).

[18] W. Weinelt, *Zeitschrift der THK* 20, 763–769 (1978).

[19] A.A. Zlotnik, *Vychisl. Protsessy Sist.* 8, 116–167 (1991).

2nd Int. Coll. on Numerical Analysis, pp. 113-122
D. Bainov and V. Covachev (Eds)

Real Zeros of Holomorphic Functions with Parameters

Joji KAJIWARA
Department of Mathematics, Kyushu University 33,
Fukuoka 812, Japan

Abstract. We seek zeros of a function $f(z,t)$, continuous in (z,t) and holomorphic in a complex variable z, and identify a parameter t_{realzero} in an interval T of the real line so that the holomorphic function $f(z,t_{\mathrm{realzero}})$ has a real zero z_{realzero}.

Keywords: Real Zeros, Holomorphic Functions with Parameter, Complex Integration.

1. Introduction

Historic Flutter Analysis of Oscillating Cascade of Jet Engine in Aeronautics, represented by K.-C. Yang, N. Yamasaki and M. Namba [4] , is done by the Finite Element Method of partial differential equations. Besides these orthodox methods, there is another access to it, Method of Parameter Identification.

Let $f(z,t)$ be a complex valued continuous function on an open set Ω in a product space $\mathbf{C} \times T$ of the complex plane $\mathbf{C}$ and a Hausdorff space T such that, for any $t \in T$, the function $f(z,t)$ of the one complex variable z is holomorphic. The said method of parameter identification starts from to identify the parameter $t_{\mathrm{realzero}} \in T$ so that the holomorphic function $f(t_{\mathrm{realzero}}, z)$ has a real zero z_{realzero} for this value t_{realzero} of the parameter $t \in T$.

In the first paper, J. Kajiwara-H. Watanabe [3] investigated this problem and showed the numerical method using Mathematica as Software and Personal Computer Macintosh II(cx and fx) as Hardware. Since the built-in Mathematics in *Mathematica* are used and since there are no theoretical errors in this method, the errors are not discussed in it. In the second paper, J. Kajiwara-M. Tsuji[2] show the numerical method using Fortran77 as Software and as Hardware, at first Personal Computer

NEC9801 for the sake of finding bugs in the programs and improvement of them, and, finally Super Computer FACOM VP-2600 for execution of numerical calculation of this identification.

The characteristic of our programs are to let the computer estimate errors by herself and seek the smallest degree p for Taylor expansion of the target holomorphic function, and the dividing number n ocurring in the numerical integration by Simpson's method etc. by herself automatically so as to obtain the sufficient accuracy.

Of course, when $f(z,t)$ is a polynomial

$$f(z,t) = \sum_{j=0}^{n} c_j(t) z^{n-j} \tag{1}$$

in z, we can solve the simutaneous algebraic equations

$$\sum_{j=0}^{n} a_j(t) x^{n-j} = 0, \sum_{j=0}^{n} b_j(t) x^{n-j} = 0 \tag{2}$$

, where $a_j(t)$ and $b_j(t)$ are, respectively, the real and imaginary part of $c_j(t)$, by the method of Kowa SEKI [Method of Laying Down Solutions, 1683] which is the Sylvester(1814→1897)'s elimination method in Europe: The parameter t_{realzero} is a zero of the Sylvester's resulvent Sylvester(t). By Mathematica, we obtain easily solutions t_{realzero} of the algebraic eqution Sylvester(t) = 0 by typing NSolve[Sylvester[t] == 0, t] and pressing the enter key. For these $t = t_{\text{realzero}}$, real zeros of the holomorphic function $f(z,t)$ are zeros which are numerically real between zeros obtained by typing NSolve[f[z,t_{realzero}] == 0, z] and pressing the enter key.

There is following absurd superstitions in Japan:

(1) It is stupid to do numerical computation by *Mathematica.*

(2) It is not wise to do numerical computation by the use of the function theory.

(3) Japanese Hardwares are more excellent than those of other developped contries.

I dare do the said parameter identification using the software Mathematica, the Hardware Macintosh and the residue theorem in the function theorey when f is a transcendental function in z in order to lay computers suitable burdens.

The aim of this work is to challenge to the above Japanese superstitions (1), (2) and (3).

2. Residue Theorem and Real Zeros

Let T be a Hausdorff space, Ω be an open set in a product space $\mathbf{C} \times T$ of the complex plane $\mathbf{C}$ and the space T, $f(z,t)$ be a complex valued continuous function on the open set Ω such that, for any $t\in T$, the function $f(z,t)$ of the one complex variable z is holomorphic in the cut $\Omega(t) = \{z\in \mathbf{C};\ (z,t)\in\Omega\} \subset \mathbf{C}$. Let q be a non-negative integer. Assume moreover that, for any $t\in T$, there exists a bounded subdomain $D(t)$ of the cut $\Omega(t)$ who commutes with the real axis $\mathbf{R}$ and whose boundary $\partial D(t)$ consists of $q+1$ piecewise smooth closed Jordan curves and the set $D = \{(z,t)\in\Omega;\ z\in D(t),\ t\in T\}$ forms a domain in the open set Ω. We define a subset

$$T_{\text{neverzero}} = \{t\in T\ ;f(z,t) \neq 0, \text{ for any } z\in\partial D(t)\} \tag{3}$$

which is open in the Hausdorff space T. For any $t\in T_{\text{neverzero}}$, we define functions

$$n(t\,;f) = \frac{1}{2\pi i}\int_{\partial D(t)} \frac{\frac{\partial f}{\partial z}(z,t)}{f(z,t)}\,dz \tag{4}$$

and

$$Z(t\,;j,f) = \frac{1}{2\pi i}\int_{\partial D(t)} z^j\,\frac{\frac{\partial f}{\partial z}(z,t)}{f(z,t)}\,dz \tag{5}$$

for any non-negative integer j, which are continuous in the open set $T_{\text{neverzero}}$ and $n(t;f)$ coincides with $Z(t;j,f)$, where $\partial D(t)$ is equipped with the usual orientatiion. By the Argument Principle of the Function Theory, the function $n(t;f)$ expresses the number of zeros of the holomorphic function $f(z,t)$ in the domain $D(t)$, the order of the zeros being counting up. More generally, each function $Z(t;j,f)$ expresses the number of the j-th power of zeros of the holomorphic function $f(z,t)$ in the domain $D(t)$, the order of the zeros being counting up.

If $n(t;f) > 0$ numerically for some point t of T, not necessarily a point of $T_{\text{neverzero}}$, then there exists zeros in the closure $\overline{D(t)} = D(t) \cup \partial D(t)$ of the domain $D(t)$. When there exists a simple pole on the boundary $\partial D(t)$ of the domain $D(t)$ for $t\notin T_{\text{neverzero}}$, the software Mathematica returns the principal value of the complex integration (5) with accuracy reports of rank 2 explained at page 14 of [14].

If $n(t) > 0$ numerically for some ponit t of $T_{\text{neverzero}}$, there exist zeros in the domain $D(t)$ for this parameter t of $T_{\text{neverzero}}$ and the number of zeros is precisely $n(t;f)$. We calculete numerically the complex integrations $Z(t;j,f)$ for $j=0,\ 1,\ 2,\ \ldots,\ n(t;f)$. Since the $n(t)$ zeros are solutions of an algeraic equation of order

$n(t; f)$ whose coefficients are expessed by the above $Z(t; j, f)$'s calculated.

For example, we assume that the Hausdorff space T is arcwise connected, $n(t; f) = 1$ in a connected open subset T_1 of the space T and that we find a pair of points t_p and t_n of the open set the domain T_1 such that the imaginary part of $Z(t_p; j, f)$ is positive and the imaginary part of $Z(t_n; j, f)$ is negative. Since T is arcwise connected, there exists a continous mapping γ of the closed interval [0,1] of the real line $\mathbf{R}$ in the domain T_1 with $\gamma(0) = t_p$ and $\gamma(1) = t_n$. Thus, without loss of generality, we may assume that the Hausdorf space T is the closed interval [0,1]. By the intermediate value theorem based on the Dedekind's axiom concerning the connectivity of the real line $\mathbf{R}$, there exists a point t_{realzero} of [0,1] such that the imaginary part of $Z(t_{\mathrm{realzero}}; 1, f)$ is zero. Then $z_{\mathrm{realzero}} = Z(t_{\mathrm{realzero}}; 1, f)$ is a real zero in the domain $D(t_{\mathrm{realzero}})$ of the holomorphic function $f(t_{\mathrm{realzero}}, z)$. Hence we can identify the parameter t_{realzero} using the residue theorem and executing complex numerical integration. When we know the convexity of the function $g_1(t)$ by an empirical knowledge, we can execute the inverse linear interpolations below. In general case, we execute the Cantor's dichotomy.

A merit of this function theoretical method is non existence of theoretical errors, because it is based on the residue theorem. In case that $n(t) > 1$, we solve numerically algebraic equation of order $n(t)$ when we calculate $n(t)$ zeros.

3. Methods of inverse linear interpolation and of bisection

In this paragraph, we assume that $T=[\alpha, \beta]$, $n(t) = 1$ for any $t \in T$ that $\mathrm{Im}\, Z(\alpha; 1, f) > 0$, $\mathrm{Im}\, Z(\beta; 1, f) < 0$, where Im stands for the imaginary part.

At first, we assume that $\mathrm{Im}\, Z(t; 1, f)$ is convex in T. For the plane $\mathbf{R}^2$, we take t and u, respectively, as the axis of abscissa and ordinate. The curve $u = \mathrm{Im}\, Z(t; 1, f)$ $(\alpha \le t \le \beta)$ is convex below. Hence the segment L which joins two points $(\alpha, \mathrm{Im}\, Z(\alpha; 1, f))$ and $(\beta, \mathrm{Im}\, Z(\beta; 1, f))$ lies under the curve $u = \mathrm{Im}\, Z(t; 1, f)$. Let $(t_1, 0)$ be the point of intersection of L and the axis of abscissa . We have

$$t_1 = \frac{\mathrm{Im}\, Z(\beta; 1, f)\, \alpha - \mathrm{Im}\, Z(\alpha; 1, f)\, \beta}{\mathrm{Im}\, Z(\beta; 1, f) - \mathrm{Im} Z(\alpha; 1, f)} \tag{6}$$

For any integer $n>0$, we define inductively

$$t_n = \frac{\mathrm{Im}\, Z(t_{n-1}; 1, f)\, \alpha - \mathrm{Im}\, Z(\alpha; 1, f)\, t_{n-1}}{\mathrm{Im}\, Z(t_{n-1}; 1, f) - \mathrm{Im}\, Z(\alpha; 1, f)} \tag{7}$$

We have

$$t_{n+1} < t_n, \ \mathrm{Im}(\mathrm{g1}(t_n)) > 0 \tag{8}$$

for any $n>0$. The monotone decreasing sequence $\{t_n;\ n=1, 2, 3, \ldots\}$ converges to a point which we denote by t_{realzero}. We put $z_{\text{realzero}} = Z(t_{\text{realzero}};1,f)$. Then z_{realzero} is a real zero of the holomorphic function $f(z, t_{\text{realzero}})$.

When we have no knowledge on the convexity of the function Im $g_1(t)$, we execute the dichotomy below. We put

$$u_1 = \alpha \ , \ v_1 = \beta \tag{9}$$

Then we have

$$\text{Im } Z(u_1;1,f) > 0 \ , \ \text{Im } Z(v_1;1,f)) < 0 \ , \ v_1 - u_1 = (\beta - \alpha)/2 \tag{10}$$

Suppose that we have taken u_k and v_k for k=1 to k=n-1 so that we have

$$\text{Im } Z(u_k;1,f) > 0 \ , \ \text{Im } Z(v_k;1,f) < 0 \ , \ v_k - u_k = (\beta - \alpha)/2^k \tag{11}$$

for k=1 to n- 1. We take the mid point

$$t_n = \frac{u_{n-1}+v_{n-1}}{2} \tag{12}$$

If Im $Z(t_n;1,f) = 0$, then this $t_{\text{realzero}}=t_n$ is the paramerter desired and $z_{\text{realzero}} = g_1(t_{\text{realzero}})$ is the real zero desired. We stop the induction finitely till n.
If Im $Z(t_n;1,f)>0$, we put $u_n=t_n$ and $v_n=v_{n-1}$. If Im $Z(t_n;1, f)<0$, we put $u_n=u_{n-1}$ and $v_n=t_n$. Then we have (11) for k=n too. In this way, we take two sequences $\{u_k; k=1, 2, 3,...\}$ and $\{v_k; k=1, 2, 3,...\}$ satisfying (11). By the Cantor 's axiom of dicotonomy, these two sequences converges to a point t_{realzero}. By (11), we have Im $g_1(t_{\text{realzero}}) = 0$. Thus we have identified the paramer $t = t_{\text{realzero}}$ so that the holomorphic function $f(z,t_{\text{realzero}})$ has a real zero $z_{\text{realzero}} = Z(t_{\text{realzero}},1,f)$ by the dichotomy.

4. Example of inverse linear interpolations by Mathematica

In order toload the machine with suitable burden, we use the Bessel function. For a non negative integer m the Bessel function of the m-th order is given by

$$J_m(z) := \sum_{k=0}^{\infty} \frac{(-1)^k z^{2k+m}}{k!(k+m)!\ 2^{2k+m}} \tag{13}$$

As in the paragraph 2.1 of the previous paper [3], we consider the function

$$f(z,t) := J_1(z + \sqrt{-1}\,(t - \frac{\pi}{10})) - J_2(z - \sqrt{-1}\,(t - \frac{\pi}{10})^2) \tag{14}$$

As the paramer space, we take the closed interval [-1,1] of the real line **R**. We seek all real zeros of the holomorphic function (14) in the open interval (1,9) of the real line **R**. In order to apply residue theorem, we take a domain containing the open interval (0,1) of the complex plane **C**. For this purpose, we take the rectangle

$$D(t) := \{z \in \mathbf{C};\ 1 < \text{Re } z < 9, -0.1 < \text{Im } z < 0.1\ \} \tag{15}$$

where Re z stands for the real part of z. We take equinoxes with mesh 0.1 in the parameter space T. For t = -1.0, the function $f(z,t)$ given by (14) and the domain $D(t)$ given by (15),we open a file of Mathematica in the Macintosh IIfx' folder and let her memorize the function $f(z,t)$ etc. typing

```
f[z_,t_]:= BesselJ[1, z+I (t-Pi)/10)] - BesselJ[2, z-I (z-Pi/10)^2],
h[z_,t_]:= D[f[z,t],z]/(2 Pi I f[z,t]),  k[z,t]:=h[z,t] z, s[z,t]:=k[z,t] z
```
(16)

and pressing the enter key. We let her calculate the complex integral $n(t\,;f)$ of (4) and $Z(t;1,f)$ of (5) by typing respectively

```
NIntegrate[h[z,t],{z,1-0.1 I,9-0.1 I,9+0.1 I,1+0.1 I,1-0.1 I],
NIntegrate[k[z,t],{z,1-0.1 I,9-0.1 I,9+0.1 I,1+0.1 I,1-0.1 I] ,
NIntegrate[s[z,t],{z,1-0.1 I,9-0.1 I,9+0.1 I,1+0.1 I,1-0.1 I] ,
```
(17)

Then the System Mathematica runs and returns values of the complex integrations (4) and (5) for j=1, 2. We list the returns of the Macintosh IIfx in the following table:

Table 1. Number of Zeros of the entire function $f(z,t)$ in the domain $D(t)$

t	calculation Time	Numerical Value of $n(t\,;f)$	Number of Zero's in $D(t)$
- 1.0	7.4 Minutes	0	0
- 0.9	7.5 Minutes	0	0
- 0.8	7.5 Minutes	0	0
- 0.7	8.3 Minutes	0	0
- 0.6	9.8 Minutes	0	0
- 0.5	10.5 Minutes	0	0
- 0.4	10.5 Minutes	0	0
- 0.3	12.9 Minutes	0	0
- 0.2	13.7 Minutes	0	0
- 0.1	18.6 Minutes	0	0
0.0	20.9 Minutes	0	0
0.1	24.2 Minutes	0 *	0
0.2	25.0 Minutes	2.00001 - 0.0000152872 I * *	2
0.3	26.7 Minutes	2. + 2.44684 10^{-11} I	2
0.4	25.1 Minutes	2. + 2.54041 10^{-9} I	2
0.5	25.0 Minutes	1.0009 - 0.000540928 I * *	1
0.6	22.7 Minutes	0.0000046528 - 0.0000155203 I *	0
0.7	22.7 Minutes	0 *	0
0.8	20.4 Minutes	0 *	0
0.9	24.2 Minutes	0 *	0
1.0	21.9 Minutes	0 *	0

The asterisk * and ** mean that Mac returns the above data with the accuracy reports of rank 1 and of rank 2, explained at page 14 of the first paper [3]. When the parameter t approaches to the boundary of $T_{\text{neverzero}}$, we receive the accuracy report and can expect to get zeros. If we compare to fishing, the accuracy reports are strikes and we can expect to fish zeros. I, who am a function theorist, love Macintosh loading with Mathematica or Theorist very much, because she attacks bravely poles of the first order and returns principal values of complex integrations, not hesitating crashes.The above table suggests us to calculate $g_1(t) := Z(t;1,f)$, $g_2(t):=Z(t;2,f)$ and, consequently, all zeros in the domain $D(t)$ from $t = 0.2$ till 0.5.

Table 2. Zeros of the entire function $f(z,t)$ in the domain $D(t)$

t	Number of zeros $n(t)$	Value of $g_1(t)$ Calculation Time	Value of $g_2(t)$ Calculation Time	Zero's of $f(z,t)$ in $D(t)$
0.2	2	8.71449+0.154908 I	43.9353+1.29214 I	2.62862+0.0858139 I
		36 Minutes	36 Minutes	6.08587+0.0690941 I
0.3	2	8.7162+0.0182089 I	43.9596+0.150667 I	2.62985+0.0102683 I
		37 Minutes	37 Minutes	6.08635+0.0079406 I
0.4	2	8.71473-0.104073 I	43.9437-0.85343 I	2.62879 -0.0597796 I
		37 Minutes	37 Minutes	6.08594 -0.0442934 I
0.5	1	6.08957-0.0907361 I		6.08957 -0.0907361 I
		30 Minutes		

In case that $g_2(t)$, there exist two zeros $z^1(t)$ and $z^2(t)$ of the entire function $f(z,t)$ in the domain $D(t)$. Since we have

$$g_1^2 = (z^1 + z^2)^2 = g_2(t) + 2z^1(t)\, z^2(t) \tag{18}$$

The two zeros $z^1(t)$ and $z^2(t)$ are two solutions of the quadratic equations

$$\zeta^2 - g_1(t)\,\zeta + \frac{g_1(t)^2 - g_2(t)}{2} = 0 \tag{19}$$

Letting Mac solve this quadratic equation, we have the above table.

Next, we do the linear inverse interpolations for the left zero putting

$$D(t) = \{z \in \mathbf{C} \,;\, 1 < \mathrm{Re}\, z < 5, -0.1 < \mathrm{Im}\, z < 0.1\},$$
$$u_1 = \alpha = 0.3, \quad v_1 = \beta = 0.4 \tag{20}$$

and obtain the following table:

Table 3. Inverse Linear Interpolations

Number of Step k	u_k / v_k / t_k	Real Part of $g_1(t)$ at u_k / at v_k	Calculation time of new $g_1(t)$	Imaginary Part of $g_1(t)$ at u_k / at v_k
1	0.3	2.62985	11.3 Minute	0.0102682
	0.4	2.62877	12.0 Minute	- 0.059788
	0.314659			
2	0.3	2.62985		0.0102682
	0.314659	2.62987	10.9 Minute	- 0.000360381

	0.314162			
3	0.3	2.62985		0.0102682
	0.314162	2.62987	11.0 Minute	- 0.00000195408
	0.314159			
4	0.3	2.62985		0.0102682
	0.314159	2.62987	11.0 Minute	- 1.05922 10^{-8}
	0.314159			
Results	0.314159	2.62987	1 Hour	0

Similary, concerning the right zero, for the domain

$$D(t) := \{z \in \mathrm{C};\ 1 < \mathrm{Re}\, z < 9, -0.1 < \mathrm{Im}\, z < 0.1\} \tag{21}$$

We have real zero 6.08636 at the parameter value $t =$ 0.314159.

5. Example for Methods of inverse linear interpolation by Fortran 77

In this paragraph, we explain the results of the second paper [2] concerning numerical method using Fortran 77 as Software and as Hardware Super Computer FACOM VP-2600 for execution of the partameter identification.

At first, we approximate the m-th order Bessel function $J_m(z)$, defined by the sum (13) for $k = 0$ to ∞, the polynomial of degree $m+2p$ given as the sum $J_{m,p}(z)$ for $k = 0$ to p, called the degree briefly. And, we approximate the entire function (14) by the polynomial $pol(t, z)$ substituting these polynomials for the Bessel functions in (14). We let the machine to seek by herself the smallest p so that the approximation error, i.e. the absolute value of $f(t, z) - pol(t, z)$ and its first derivative with respect to z, are less than 10^{-15}. We consider the complex integration along the boundary of the rectangle $\{z \in \mathrm{C}\ ;\ 2.5 < \mathrm{Re}\, z < 2.75, -0.1 < \mathrm{Im}\, z < 0.1\}$ for the polynomial $h(z)=z^j\ (\partial pol(t, z)/\partial z)/(2\pi\ \mathrm{i}\, pol(t, z))$ and represent it as the sum of four integral along the segment, forming the boundary, with respect to the real parameter s. The Simpson Error, absolute value of difference of an integral of a real-valued C^4 function $h(s)$ along the closed interval $a \le s \le b$ has, as an upper bound, the maximum of the absolute value of the fourth derivative of the function $h(s)$ on $[a,b]$ times $(b-a)^5/2880$ and this majoration is valid even when $h(s)$ is complex-valued too. Moreover, when $h(s)$ is holomorphic in the closed r-neighborhood $K(r)$ of the closed interval $[a,b]$, the maximum on the closed interval $[a,b]$ of the absolute value of the fourth derivative of the function $h(s)$ has, as an upper bound, the maximum of the absolute value of the holomorphic function $h(s)$ on the closed r-neighborhood $K(r)$ times $4!r^{-4}$. Since the $pol(t, z)$ is in the denominater, we let the machine calculate the minimum of the absolute value of the $pol(t, z)$ on the dividing point while she computes numerically on Simpson Method, in order to obtain the Error.

The characteristic of our programs are to let the machine estimate errors by herself and choose the smallest p in the approximation of $f(t, z)$ by $pol(t, z)$ so that the approximation error is smaller than 1.0×10^{-15} and the smallest interger $k(0 \leq k \leq 7)$ of the dividing numbers $N=2^k \times 100$ for numerical integration etc. automatically by herself so that, in the calculation of $Z(t\,;0, f)$, the Simpson Error is smaller than 0.5 and, in the calculation of $Z(t\,;1, f)$, the Simpson Error is smaller than the absolute value of the imaginary part $Z(t\,;1, f)$. Then, we can correctly conclude that the numbers of zeros are one and the signature of the zero is negative, she report these situatuation as Error Code 0 and otherwise Error code 1 in the following table 4. We only report the results in the following table. Please read the second paper [2].

Table 4. Return of FACOM VP-2600 for Program in FORTRAN77

Value of t	Value of N	Number of Zeros $Z(t\,;0, f)$ Simpson Error	Value of Zero Point $Z(t\,;1, f)$ Simpson Error
0.3	200	$(1.0000, 8.8\times10^{-14})$ 0.259 Error Code 0	
	800		$(2.6298, 1.02\times10^{-2})$ 2.795×10^{-3} Error Code 0
0.4	800	$(1.0000, -3.132\times10^{-15})$ 3.742×10^{-2} Error Code 0	
	1600		$(2.62877, -5.977\times10^{-2})$ 6.448×10^{-3} Error Code 0
0.314658	200	$(1.0000, -3.224\times10^{-15})$ 0.225 Error Code 0	
	1600		$(2.62987458, -3.60\times10^{-4})$ 1.520×10^{-4} Error Code 0
0.314161	200	$(1.0000, -1.747\times10^{-17})$ 0.225 Error Code 0	

	6400		(2.629874632, -1.95×10^{-6}) 5.936×10^{-7} Error Code 0
0.31415928	200	(1.0000, -9.46×10^{-20}) 0.225 Error Code 0	
	12800		(2.62987463477, -1.05×10^{-8}) 3.710×10^{-8} Error Code 1
0.314159265359	200	(1.0000, -5.124×10^{-22}) 0.225 Error Code 0	
	12800		(2.62987463477, -5.72×10^{-11}) 3.710×10^{-8} Error Code 1

Calculation Time 112 Second

Thus we have identified by the super computer FACOM VP-2600 the parameter value $t = 0.314159265$, for which the holomorphic function $f(z,t)$ has a real zero $z = 2.6298746$, using the sofware Fortran77, what proves accuracy of the results obtained by Macintosh II in the preceding paragraph, using the software *Mathematica*.

References

[1] L. Hörmander, An Introduction to Complex Analysis in Several Variables, D. Van Nostrand 1966, 208 pages.

[2] J. Kajiwara and T. Tsuji, *Real Zeros of Holomorphic Functions with Parameters* II: *Complex numerical integration using Fortran 77*, submitted to the Journal of Mathematics, Toyama Univ. (1994).

[3] J. Kajiwara and H. Watanabe, *Real Zeros of Holomorphic Functions with Parameters* I: *Complex numerical integration using Mathematica*, Mem. Fac. Sci. Kyushu Univ. **46**-1 (1992), 1-30.

[4] K.-C. Yang, N. Yamasaki and M. Namba, *Numerical Calculation of Unsteady Aerodynamic Forces for Three-Dimensional Subsonic Oscillating Cascade by a Finite Element Method*, Mem. Fac. Eng. Kyushu Univ. **48**(1988), 253-280.

Phone: +81-92-641-1101 Ext. 4353

FAX: To Prof. Kajiwara +81-92-632-2737

e-mail: kajiwara@math.sci.kyushu-u.ac.jp

2nd Int. Coll. on Numerical Analysis, pp. 123-132
D. Bainov and V. Covachev (Eds)

Optimal Partitioning
in Univariate and Multivariate Approximation

Günter Meinardus

Universität Mannheim, 68131 Mannheim, Germany

Abstract. *Uniform segmential approximations have been developed since a longer time. A few years ago an efficient numerical algorithm to compute numerically best approximations for certain types of function spaces was presented by G.Nürnberger, M.Sommer and H.Strauß. In the specialization of this procedure one was able to make use of general results concerning the approximations in small intervals and concerning analytic properties of the function to be approximated. The most interesting question is that of an a priori information on the quantitative improvement of accuracy. In this paper a number of special results and some asymptotic relations concerning the optimal location of the knots and concerning the improvement factor are presented. Then we give a theorem on optimal and leveled partitioning of rectangular and of circular domains in the bivariate case. In the latter case it is recommended to use spaces consisting of spherical harmonics for the approximation. Some special cases are investigated also in the bivariate concept in order to get some information what is to be expected concerning the improvement of accuracy.*

Keywords. Uniform approximation, segmential approximation, optimal partitioning of rectangular and circular domains.

1. Segmential Approximation. Let a closed interval $J = [a, b] \in I\!R$ be given. We consider the space $C(J)$ of real or complex functions f on $[a, b]$, furnished with the Chebyshev norm

$$\|f\| = \sup\{|f(x)| \,\big|\, x \in J\}.$$

Let V be a subspace of $C(J)$ of finite dimension. An element $v \in V$ is called a best approximation of a given function $f \in C(J)$, if the inequality

$$\|f - v\| \leq \|f - w\|$$

holds for every function $w \in V$. In this case we call

$$\rho(f) = \|f - v\|$$

the minimal deviation of f with respect to V.

For $k \in \mathbb{N},\ k \geq 2$ we assume that the interval J is divided into k subintervals

$$[\tau_0, \tau_1],\ [\tau_1, \tau_2],\ \ldots,\ [\tau_{k-1}, \tau_k]$$

with

$$a = \tau_0 \leq \tau_1 \leq \cdots \leq \tau_{k-1} \leq \tau_k = b \tag{1}$$

by some points $\tau_1,\ \tau_2,\ \ldots,\ \tau_{k-1}$. In every subinterval $J_\nu = [\tau_{\nu-1},\ \tau_\nu]$ let us be given subspaces V_ν and corresponding minimal deviations $\rho_\nu(f) = \rho(f; \tau_{\nu-1},\ \tau_\nu)$ for given functions $f \in C(J),\ \nu = 1,\ 2,\ \ldots,\ k$. The segmential approximation problem consists in finding such numbers τ_ν satisfying (1) such that the

$$\max_\nu\ \rho_\nu(f)$$

is as small as possible.

The problem was stated and investigated by Ch.L.Lawson [1] in 1964. He considered even rational approximation problems of this type. A general approach was given by D.E.McLure [2], and in [3] for almost arbitrary non-linear approximation families. The motivation for this type of approximation is of course that one hopes to gain a significant amount of accuracy, compared with the unpartitioned case.

The set of knots $\tau_1,\ \tau_2,\ \ldots,\ \tau_{\nu-1}$ satisfying (1) is referred to as a *partition*

$$T_k = \{\tau_1,\ \tau_2,\ \ldots,\ \tau_{k-1}\}. \tag{2}$$

We will now restrict to polynomial approximation, i.e. for $n \in \mathbb{N}_0$ we approximate a given function $f \in C(J)$ in every subinterval J_ν of the partition T_k by polynomials $p \in \Pi_n$. Let

$$\rho^*(f) = \inf_{T_k}\{\max_\nu\ \rho_\nu(f)\}$$

be the minimal deviation of the segmential approximation problem. Here the maximum is taken over $\nu = 1,\ 2,\ \ldots,\ k$, and the infimum is formed over all possible partitions T_k. If we are interested in the dependence of this minimal deviation on n and k, we will write $\rho^*_{n,k}(f)$ instead. Of course $\rho^*_{n,1}(f)$ denotes the minimal deviation for the undivided interval J.

Theorem 1 (Ch.L.Lawson [1]): *For every partition T_k the inclusion*

$$\min_\nu\ \rho_\nu(f) \leq \rho^*(f) \leq \max_\nu\ \rho_\nu(f) \tag{3}$$

is valid.

We call T_k a *leveled* partition, if

$$\rho_1(f) = \rho_2(f) = \cdots = \rho_k(f)$$

holds.

Theorem 2 (Ch.L.Lawson [1]): *For every $f \in C(J)$ and every value of k there exists a leveled partition.*

It follows with Theorem 1 that a leveled partition is also an optimal partition, i.e. we have in this case

$$\rho_\nu(f) = \rho^*(f), \; \nu = 1, 2, \ldots, k.$$

A best segmential approximation consists of the polynomials p_ν, $\nu = 1, 2, \ldots, k$, where p_ν is the best approximation of f on the interval $[\tau_{\nu-1}, \tau_\nu]$ with respect to the space Π_n, provided T_k is an optimal partition. In the following we denote a leveled partition by

$$T_k^* = \{\tau_1^*, \tau_2^*, \ldots, \tau_{k-1}^*\}.$$

If $f \notin \Pi_n$, then obviously

$$a < \tau_1^* < \tau_2^* < \cdots < \tau_{k-1}^* < b$$

holds true.

A method of constructing a best segmential approximation will consist of computing a leveled partition. Some of such methods have been developed many years ago, e.g. in [4], by C.DeBoor [5],[6] and by H.G.Burchard [7],[8]. But these procedures all have some disadvantages. Surely none of them can be applied to arbitrary functions f. In 1986 G.Nürnberger, M.Sommer and H.Strauß [9] presented an algorithm, which converges for every function $f \in C(J) \setminus \Pi_n$. We will not describe it here, but refer to the papers [9] and [10] instead. Generalizations to splines with free knots can be found in G.Nürnberger [11] and [12].

2. The Segmented Minimal Deviation. If f belongs to some subspace of $C(J)$, it is possible to give a priori estimates of the segmented minimal deviation $\rho_{n,k}^*(f)$. The proof of the following assertion will be given in a forthcoming paper.

Theorem 3 : *Let $f \in C^{(n+1)}(J)$. Then there exists a number $\xi^* \in J$, depending on f, n and k such that the equality*

$$\rho_{n,k}^*(f) = \frac{|f^{(n+1)}(\xi^*)|}{2^{2n+1}(n+1)!\, k^{n+1}} (b-a)^{n+1} \tag{4}$$

holds true.

Hence in many cases we will gain approximately a factor k^{n+1} in accuracy, comparing the undivided case with the k-fold segmentation. This remark agrees asymptotically with

Theorem 4 (G.M.Phillips [13]): *Let $f \in C^{(n+1)}(J)$. Furthermore we assume that there is no interval $[\alpha, \beta]$ with $a \leq \alpha < \beta \leq b$ on which the restriction of f coincides with a polynomial of the space Π_n. Then*

$$\rho_{n,k}^*(f) = \frac{\Lambda_n(f)}{2^{2n+1}(n+1)!\, k^{n+1}} \; (1 + o(1))$$

for $k \to \infty$. *Here*

$$\Lambda_n(f) = \left(\int_a^b |f^{(n+1)}(t)|^{\frac{1}{n+1}} dt \right)^{n+1}.$$

A generalization, concerning a density function of the knots of a leveled partition, has been proved by R.Grothmann a.H.N.Mashkar [14].

We present now a collection of simple examples, which can be treated theoretically.

Example 1. Let $J = [0,\ 1]$ and $f(t) = t^{n+1}$. Using the best approximation property of the Chebyshev polynomials, we get easily for any pair x, y with $x < y$ the relation

$$\rho(f; x, y) = \frac{(y-x)^{n+1}}{2^{2n+1}}.$$

In order to construct a leveled partition, the minimal deviations

$$\rho_\nu(f) = \rho(f; \tau_{\nu-1}, \tau_\nu) = \frac{(\tau_\nu - \tau_{\nu-1})^{n+1}}{2^{2n+1}}$$

have to be equal for $\nu = 1, 2, \cdots, k$. Hence, the only leveled partition ist the equidistant one, i.e.

$$\tau_\nu^* = \frac{\nu}{k}, \quad \nu = 0,\ 1,\ \ldots,\ k.$$

Furthermore we have

$$\rho_{n,k}^* = \frac{1}{k^{n+1} 2^{2n+1}}.$$

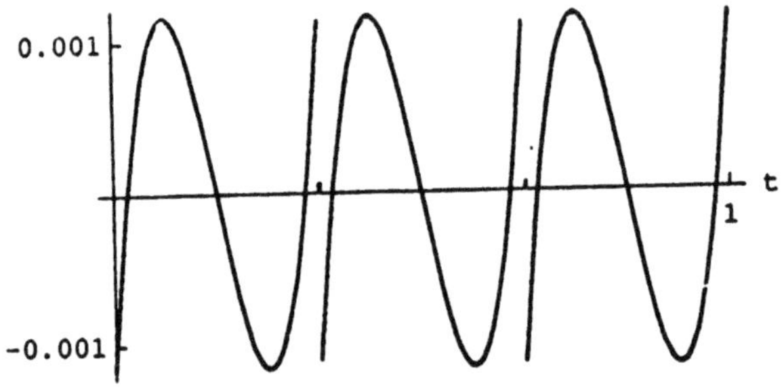

Fig.1: Error in example 1 (n=1, k=3).

Example 2. Let $J = [0,\ 1]$, $f(t) = \sqrt{t}$ and $n = 1$. The best approximation $p \in \Pi_1$ to f on the interval $[x,\ y]$ with $0 \le x < y \le 1$ can be computed easily to be

$$p(t) = \frac{t}{\sqrt{x} + \sqrt{y}} + \frac{x + 6\sqrt{xy} + y}{8(\sqrt{x} + \sqrt{y})}.$$

The minimal deviation turns out to be

$$\rho(f; x, y) = \frac{1}{8} \frac{(\sqrt{y} - \sqrt{x})^2}{\sqrt{x} + \sqrt{y}}.$$

In order to construct a leveled partition we put $z_\nu = \sqrt{\tau_\nu}$ for $\nu = 0, 1, \ldots, k$. Then we have to solve the system of equations

$$\frac{(z_\nu - z_{\nu-1})^2}{z_\nu + z_{\nu-1}} = \frac{(z_{\nu+1} - z_\nu)^2}{z_{\nu+1} + z_\nu}$$

for $\nu = 1, 2, \ldots, k$. A short computation yields the unique solution

$$z_\nu = \frac{\nu(\nu+1)}{k(k+1)},$$

i.e.

$$\tau_\nu^* = \left(\frac{\nu(\nu+1)}{k(k+1)}\right)^2$$

for $\nu = 0, 1, \ldots, k$. Furthermore we get

$$\rho_{1,k}^*(f) = \frac{1}{4k(k+1)}.$$

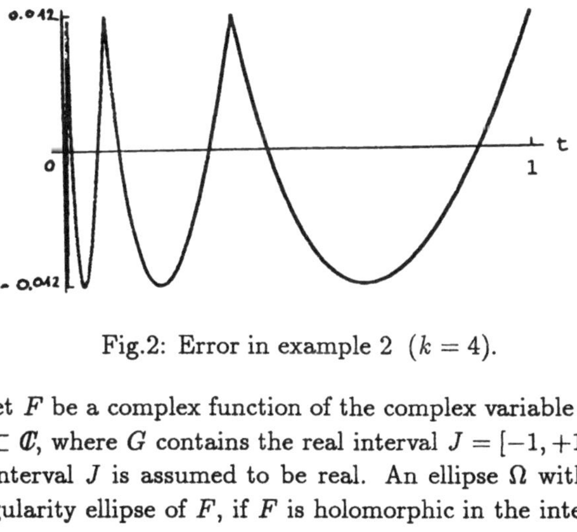

Fig.2: Error in example 2 $(k = 4)$.

Example 3. Let F be a complex function of the complex variable z, holomorphic in some region $G \subset \mathbb{C}$, where G contains the real interval $J = [-1, +1]$. The restriction f of F to the interval J is assumed to be real. An ellipse Ω with foci $+1$ and -1 is called the regularity ellipse of F, if F is holomorphic in the interior of Ω and not in any larger ellipse with the same foci. Let F possess a finite regularity ellipse Ω_c, where we have denoted by c the real point larger than 1 on the circumference of Ω_c, $c = c(f)$. We put

$$q = q(f) = c + \sqrt{c^2 - 1}.$$

It is well known that the minimal deviation $E_n(f)$ in approximating the function f by polynomials from the space Π_n behaves geometrically, i.e.

$$\limsup(E_n(f))^{\frac{1}{n}} = \frac{1}{q}.$$

If one considers any interval $[x, y]$ with $-1 \leq x < y \leq 1$, then, in order to compute the corresponding geometric factor, one has to look for the largest ellipse Ω_c with foci x and y such that F is holomorphic in its interior, and transform the interval by translation and scaling to the interval $[-1,+1]$.

We will restrict to an important special case now, making the following assumptions:
a) The function F possesses a singularity at the point $z = c$.
b) The function F is holomorphic in the disk

$$|z+1| \leq 1 + c.$$

It follows that the minimal deviation $E_n(f; x.y)$ in approximating the function f by polynomials from the space Π_n on the interval $[x, y]$ behaves geometrically with the parameter

$$c(x,y) = \frac{2c - x - y}{y - x}$$

instead of c, and with the geometric factor

$$q(f; x, y) = c(x,y) + \sqrt{\left(c(x,y)\right)^2 - 1}$$

instead of q.

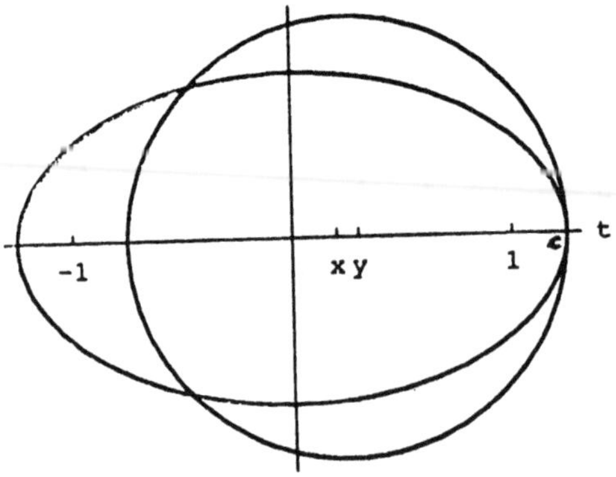

Fig.3: The ellipsis in example 3.

Hence, in order to get asymptotically, for $n \to \infty$, a leveled partition, the intervals $J_\nu = [\tau_{\nu-1}, \tau_\nu]$ all have to produce the same parameters, i.e. the equations

$$c(\tau_{\nu-1}, \tau_\nu) = c(\tau_\nu, \tau_{\nu+1})$$

have to be satisfied for $\nu = 1, 2, \ldots, k$. It is easy to verify that this system of equations possesses the only solution

$$\tau_\nu^* = c - (c+1)\left(\frac{c-1}{c+1}\right)^{\frac{\nu}{k}}$$

for $\nu = 0, 1, \ldots, k$. The associated parameter resp. geometric factor are

$$c_k^* = \frac{1 + \left(\frac{c-1}{c+1}\right)^{\frac{1}{k}}}{1 - \left(\frac{c-1}{c+1}\right)^{\frac{1}{k}}}$$

and

$$q_k^* = c_k^* + \sqrt{c_k^{*2} - 1}.$$

If F is an entire function, the leveled partition converges for $n \to \infty$ to the equidistant distribution.

3. Optimal Partitioning of a Rectangular Domain. In his investigations concerning H-sets L.Collatz [15] recommended to consider the following bivariate problem: Let us be given a rectangle $\Delta \subset \mathbb{R}^2$,

$$a \leq u \leq b, \quad c \leq v \leq d.$$

For $r, k \in \mathbb{N}$, $r \geq 2$, $k \geq 2$ let the region Δ be partitioned into r strips J_μ defined by the inequalities

$$c = \tau_0 \leq \tau_1 \leq \cdots \leq \tau_{r-1} \leq \tau_r = d,$$

where

$$J_\mu = \{(u,v) \in \Delta | \; a \leq u \leq b \,; \; \tau_{\mu-1} \leq v \leq \tau_\mu\},$$

for $\mu = 1, 2, \ldots, r$. Furthermore each strip is partitioned into k smaller rectangles $\Delta_{\mu,\nu}$ defined by the inequalities

$$a = \sigma_{\mu,0} \leq \sigma_{\mu,1} \leq \cdots \leq \sigma_{\mu,k-1} \leq \sigma_{\mu,k} = b,$$

where

$$\Delta_{\mu,\nu} = \{(u,v) \in \Delta | \; \sigma_{\mu,\nu-1} \leq u \leq \sigma_{\mu,\nu}; \; \tau_{\mu-1} \leq v \leq \tau_\mu\}$$

for $\mu = 1, 2, \ldots, r$; $\nu = 1, 2, \ldots, k$.

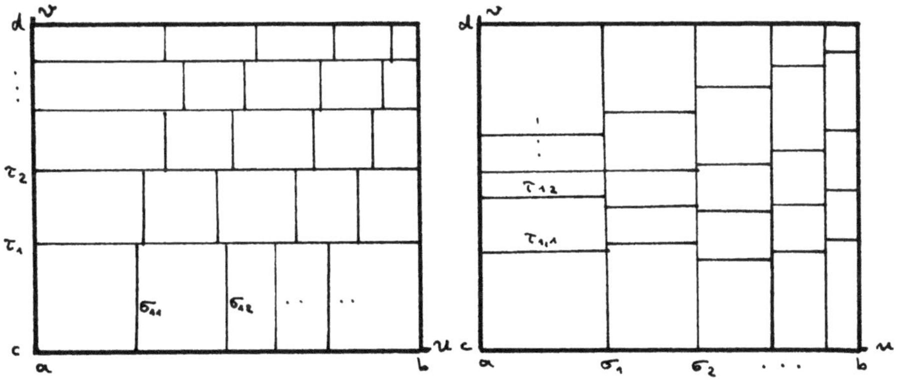

Fig.4: Partitioning a rectangle.

The partitioned approximation problem on Δ for a given function $f \in C(\Delta)$ with respect to a certain finite dimensional vector space V (e.g. some polynomial spaces in two variables) is as follows:

Find a partition of Δ of the above type into rectangles such that the expression

$$\rho_{V,k,r}(f) = \max_{\mu,\nu} \rho_{\mu,\nu}(f)$$

is as small as possible. Here $\rho_{\mu,\nu}(f)$ denotes the minimal deviation in approximating the function f by functions from the space V on the rectangle $\Delta_{\mu,\nu}$ in the uniform sense.

In a forthcoming joint paper with G.Nürnberger we will present results on the inclusion of the partitioned minimal deviation and on the existence of a leveled partition. As a matter of fact one has to equalize rk minimal deviations, and there are exactly $rk - 1$ free parameters to do this.

Furthermore we will generalize the construction algorithm for a leveled partition and prove its overall convergence.The simplifications in order to accelerate the speed of convergence of this algorithm will be investigated also. Here the basic idea consists in using generalizations of the Chebyshev polynomials to the bivariate case and then constructing simple linear functionals, which in modulus are good substitutes for the minimal deviations. The latter remark makes sense only, if V is a suitable polynomial space, For details cp. H.Ehlich and K.Zeller [16]. Of course one may pose the problem by interchanging the variables u and v also.

Generalizations of this approach to multivariate problems with respect to hypercubes are evident.

4. Optimal Partitioning of a Disk. The idea for rectangles can be carried over in two ways of partitioning a disk:
a) The disk is partitioned into a smaller disk and $r - 1$ annuli ; then every annulus and the small disk are divided into k segments resp. sectors.
b) The disk is partitioned into r sectors ; then every sector is divided into k segments by parts of concentric circles.

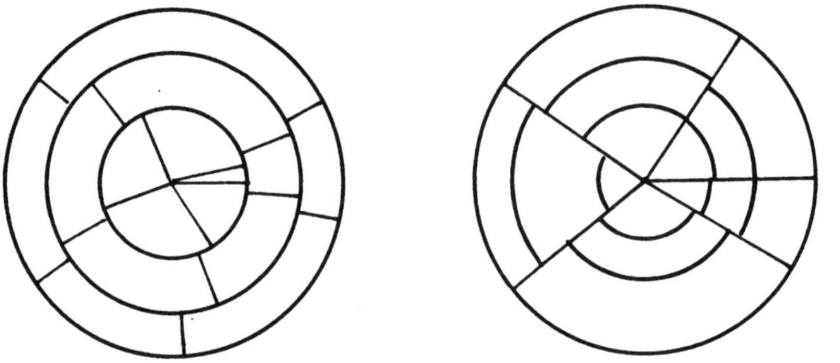

Fig.5: Partitioning of a disk.

Again the optimal partitioning problem consists in finding a partition of this pattern such that the maximum of all the minimal deviations in approximating a given continuous function on each of the small regions with respect to some finite dimensional vector space is as small as possible. It seems reasonable to consider as the approximating space (in polar coordinates)

$$V = span\ \{1, r\ cos(\phi),\ r\ sin(\phi), \ldots,\ r^n cos(n\phi),\ r^n sin(n\phi)\}$$

for $n \in \mathbb{N}_0$. In simplifying the numerical procedure one may use a connection with complex polynomial approximation on sectors etc. For details cp. G.Geiger a.G.Opfer [17] and U.Grothkopf a.G.Opfer [18]. In generalizations to the multivariate case concerning balls in $\mathbb{R}^d$ it is recommended to use spaces of spherical harmonics.

References.

[1] Ch.L.Lawson: Characteristic Properties of the Segmented Rational Minmax Approximation Problem. Num.Math.**6**, 293-301 (1964).

[2] D.E.McLure: Nonlinear Segmented Function Approximation and Analysis of Line Patterns. Report Brown University, Providence (1973).

[3] G.Meinardus: Approximation of Functions: Theory and Numerical Methods. Springer-Verlag Heidelberg (1967).

[4] G.Meinardus: Zur Segmentapproximation mit Polynomen. ZAMM **46**, 239-246, (1966).

[5] C.DeBoor: Good Approximations by Splines with Variable Knots. In: Spline Functions and Approximation Theory (eds. A.Meir, A.Sharma), Birkhäuser-Verlag Basel, 57-72 (1973).

[6] C.DeBoor: Good Approximations by Splines with Variable Knots II. In: Numerical Solutions of Differential Equations (ed.G.A.Watson), Lect. Notes in Math. **363**, 12-20 (1974).

[7] H.G.Burchard: Splines (with Optimal Knots) are Better. Applicable Analysis **3**, 309-319 (1974).

[8] H.G.Burchard: Piecewise Polynomial Approximation on Optimal Meshes. Journ. Appr.Th. **14**, 128-147 (1975.

[9] G.Nürnberger, M.Sommer a.H.Strauß: An Algorithm for Segment Approximation. Num.Math **48**, 463-477 (1986).

[10] G.Meinardus, G.Nürnberger, M.Sommer a.H.Strauß: Algorithms for Piecewise Polynomials and Splines with Free Knots. Math.of Comp. **53**, 235-247 (1989).

[11] G.Nürnberger: Approximation by Univariate and Bivariate Splines. These Proceedings.

[12] G.Nürnberger: Approximation by Spline Functions. Springer-Verlag Heidelberg (1989).

[13] G.M.Phillips: Error Estimates for Best Polynomial Approximations. In: Approximation Theory (ed.A.Talbot), Acad.Press London, 1-6 (1970).

[14] R.Grothmann a.H.N.Mhaskar: Detection of Singularities Using Segment Approximation. Math.of Comp. **59**, 533-540 (1992).
[15] L.Collatz: Einschließungssatz für die Minimalabweichung bei der Segmentapproximation. In: Simp.Intern.sulle Applic.Analisi alla Fisica Matematica, 11-21 (1964).
[16] H.Ehlich a.K.Zeller: Čebyšev-Polynome in mehreren Veränderlichen. Math.Z. **93**,142-143 (1966).
[17] C.Geiger a.G.Opfer: Complex Chebyshev Polynomials on Circular Sectors. Journ.Appr.Th. **24**,93-118 (1978).
[18] U.Grothkopf a.G.Opfer: Complex Chebyshev Polynomials on Circular Sectors with Degree six or less. Math.of Comp. **39**, 599-615 (1982).

2nd Int. Coll. on Numerical Analysis, pp. 133-142
D. Bainov and V. Covachev (Eds.)

Asymptotic Behavior of Blow-Up Solutions of the Nonlinear Schrödinger Equation with Critical Power Nonlinearity

Hayato NAWA
Department of Mathematics, Faculty of Science, Tokyo Institute of Technology
2-12-1 oh-okayama meguro, Tokyo 152, Japan

Abstract. This article is a collection of author's results of recent papers [23–28] concerning the blow-up problem for the nonlinear Schrödinger equation with critical power nonlinearity on $\mathbb{R}^N$, in which it is shown that every blow-up solution concentrates its L^2 mass at most finite points in some sense.

1. INTRODUCTION

This article is a collection of author's results of recent papers [23–28] concerning the nonlinear Schrödinger equation with critical power nonlinearity (see (NSC) below) where the idea of concentrated compactness or of bubble theorems have been emphasized. Our main results are Theorem A (Sect 1), Theorem B (Sect 2) and Theorem C (Sect 3).

We start with a review of the Cauchy problem for the nonlinear Schrödinger equation:

$$\mathrm{C(p)} \qquad \begin{cases} \text{(NS)} \quad 2i\dfrac{\partial u}{\partial t} + \Delta u + |u|^{p-1}u = 0, & (t,x) \in \mathbb{R}_+ \times \mathbb{R}^N, \\ \text{(IV)} \quad u(0,x) = u_0(x), & x \in \mathbb{R}^N. \end{cases}$$

Here $i = \sqrt{-1}$, $u_0 \in H^1(\mathbb{R}^N)$ and Δ is the Laplace operator on $\mathbb{R}^N$. The unique local existence of solutions of C(p) is well known for $1 < p < 2^* - 1$ ($2^* = \frac{2N}{N-2}$ if $N \geqq 3$, $= \infty$ if $N = 1, 2$): For any $u_0 \in H^1(\mathbb{R}^N)$, there exist a unique solution $u(t,x)$ of C(p) in $C([0,T_m); H^1(\mathbb{R}^N))$ for some $T_m \in (0,\infty]$ (maximal existence time), and $u(t)$ satisfies the following two conservation laws of L^2 and the energy:

$$\|u(t)\| = \|u_0\|, \tag{1.1}$$

$$E_{p+1}(u(t)) \equiv \|\nabla u(t)\|^2 - \frac{2}{p+1}\|u(t)\|_{p+1}^{p+1} = E_{p+1}(u_0), \tag{1.2}$$

for $t \in [0,T_m)$, where $\|\cdot\|$ and $\|\cdot\|_{p+1}$ denotes the L^2 norm and L^{p+1} norm respectively. Furthermore $T_m = \infty$ or $T_m < \infty$ and $\lim_{t\to T_m} \|\nabla u(t)\| = \infty$. For details, see, *e.g.*, [11,12,14].

As for the existence and non-existence of global solutions of C(p), the following is well known (see [11,12,13,14,30,31]).

(i) If $1 < p < 1 + \frac{4}{N}$, there exists a global solution $u \in C_b(\mathbb{R}; H^1(\mathbb{R}^N))$, for any $u_0 \in H^1(\mathbb{R}^N)$, where $C_b(\mathbb{R}; H^1(\mathbb{R}^N)) = C(\mathbb{R}; H^1(\mathbb{R}^N)) \cap L^\infty(\mathbb{R}; H^1(\mathbb{R}^N))$.

(ii) If $1 + \frac{4}{N} \leqq p < 2^* - 1$, there is a subset $\mathcal{B} \in H^1(\mathbb{R}^N)$ such that for any $u_0 \in \mathcal{B}$ the solution of C(p) blows up, *i.e.* the L^2 norm of its gradient explodes in finite time T_m.

As we have seen above, the number $p = 1 + \frac{4}{N}$ is the critical number for the existence of blow-up solutions of C(p). In what follows, we refer to (NS) with $p = 1 + \frac{4}{N}$ as (NSC), *i.e.*,

$$\text{(NSC)} \qquad 2i\frac{\partial u}{\partial t} + \Delta u + |u|^{\frac{4}{N}} u = 0,$$

and we shall use the notations

$$E(v) = E_\sigma(v), \qquad \sigma = 2 + \frac{4}{N}.$$

The nonlinear Schrödinger equation of the form (NSC) is of physical interest, because (NSC) with $N = 2$ arises in a theory of the stationary self-focusig of a laser beam propagating along the t-axis in a nonlinear medium (see *i.e.* [1,2,15,40]). We may say that the blow-up of solution corresponds to the focusing of the laser beam.

Recently, many mathematicians have studied the formation of singularities in blow-up solutions of (NSC) near blow-up time (*e.g.*, [7, 15, 19, 20, 21, 22, 23–29, 35, 37, 38]). Here, it is worth while to note that (NSC) has a remarkable property that it is invariant under the pseudo-conformal transformations (see *e.g.*,[7,29] and (2.2) below). This symmetry seems to be closely related to the structure of solutions of (NSC) (see *e.g.*, Weinstein [37,38], Nawa and M. Tsutsumi [29] and Cazenave and Weisler [7]): in the super critical case ($p > 1 + \frac{4}{N}$), Merle [19] suggested that every blow-up solution of (NS) has a strong limit in L^2 at blow-up time; in the critical case ($p = 1 + \frac{4}{N}$), Nawa [23,24] and Weinstein [38] showed that every blow-up solution of (NSC) loses its L^2 continuity at blow-up time because of the concentration of its L^2 mass (see also Merle [20], Merle and Y. Tsutsumi [21], and Y. Tsutsumi [35]). Moreover we know how amount the blow-up solution of (NSC) concentrate their L^2 mass. Precisely, we can prove [23,24,26]

Theorem A. *Let Q be a nontrivial solution of the elliptic equation*

$$\text{(A.1)} \qquad \Delta Q - Q + |Q|^{\frac{4}{N}} Q = 0$$

such that

$$\text{(A.2)} \qquad \frac{2}{\sigma}\|Q\|^{\frac{4}{N}} = \inf_{\substack{v \in H^1(\mathbf{R}^N) \\ v \not\equiv 0}} \frac{\|v\|^{\frac{4}{N}} \|\nabla v\|^2}{\|v\|_\sigma^\sigma}$$

$$= \inf_{\substack{v \in H^1(\mathbf{R}^N) \\ v \not\equiv 0}} \left\{ \frac{2}{\sigma}\|v\|^{\frac{4}{N}} \; ; \; \|\nabla u(t)\|^2 - \frac{2}{\sigma}\|u(t)\|_\sigma^\sigma \leqq 0 \right\}.$$

Let $u(t)$ be the singular solution of (NSC) such that

$$\text{(A.3)} \qquad \lim_{t \to T_m} \|\nabla u(t)\| = \lim_{t \to T_m} \|u(t)\|_\sigma = \infty$$

for some $T_m \in (0, \infty]$. We put

$$\text{(A.4)} \qquad \lambda(t) = \frac{1}{\|u(t)\|_\sigma^{\sigma/2}},$$

$$\text{(A.5)} \qquad u_\lambda(t, x) = \lambda(t)^{\frac{N}{2}} u(t, \lambda(t)x),$$

$$\text{(A.6)} \qquad A = \sup_{R>0}\left(\liminf_{t \uparrow T_m} \left(\sup_{y \in \mathbf{R}^N} \int_{|x-y| \leqq R} |u_\lambda(t,x)|^2 dx \right)\right).$$

(1) Then we have

$$A \geqq \|Q\|^2 \tag{A.7}$$

and we have that, for any $\varepsilon \in (0,1)$, there are constans $K > 0$ and $T_0 > 0$, and there is a right continuous function $\gamma(\cdot) \in L^\infty_{loc}([T_0, T_m); \mathbb{R}^N)$ such that

$$\int_{|x-y| \leqq R\lambda(t)} |u(t, x + \lambda(t)\gamma(t))|^2 dx \geqq (1-\varepsilon)A, \quad t \in [T_0, T_m), \tag{A.8}$$

for any $R \geqq K$. Moreover if $A > \frac{1}{2}\|u(0)\|^2$, then we have $\gamma(\cdot) \in C([T_0, T_m); \mathbb{R}^N)$.
(2) Suppose in addition that $\|u(0)\|^2 < 2\|Q\|^2$. Then by (1), we have $\gamma(\cdot) \in C([T_0, T_m); \mathbb{R}^N)$.

Remark 1.1. (1) The equation (A.1) is a time-independent version of (NSC) and arises in various domain of physics. See [3,6,27,28,32,36] for the existence of positive solutions of (A.2) and for the associated minimization problems. The standard argument shows that $Q \in \mathcal{S}$ (the space of C^∞ functions of rapid decreasing). We can also prove that $E(Q) = 0$.
(2) By the first equality in (A.2) and the conservation law (1.2), we see that if $\|u_0\| < \|Q\|$, the corresponding solution exists globally in time. For this, see Weinstein [36, 37]. Furthermore the initail datum $Q(x)e^{-i|x|^2/2}$ leads to the blow-up solution which blows up at time $t = 1$ (see [29,37]). In this sense, the estimate (A.7) is optimal.

The formula (A.8) tells us that the focusing of a laser beam could be understood mathematically as "L^2 concentration" phenomena of blow-up solutions of (NSC). The quantity A measures the "size" of the "largest" singularity, since the blow-up solution, in genaral, has several L^2-concentration points (see Merle [20] and Theorem B in Sect.2 of this paper).

It is worth while to note here taht the nonlinear Schrödinger equation (NS) also derived from a field equation for a quantum mechanical nonrelativistic *many body system* in the semi classical limit (*e.g.* [11] and the talk of Professor J. Schmeelk delivered at the 4th International Colloquium on Differential Equations). Hence it is expected that the solution of nonlinear Schrödinger equation involves a *many body* character in its behavior. Therefore, in general, it seems difficult to trace directly the dynamics of blow-up solutions. In other words, we may not expect the clear "trajectory"of singularities as in Theorem A (2) generally.

In the next section, we state our theorem concerning the asymptotic behavior of blow-up solutions of (NSC).

2. ASYMPTOTIC OF BLOW-UP

Before stating our main theorem, we recall some previous results as to the asymptotic behavior of blow-up solution of (NSC).

(I) Let $u(t)$ be a solution of (NSC) such that $\|u(t)\| = \|Q\|$ and $\|\nabla u(t)\| \to \infty$ as $t \to T_m$ for some $T_m \in (0, \infty]$. Then we have, for $\tilde{\lambda}(t) = \|\nabla u(t)\|^{-1}$,

$$\left\| \tilde{\lambda}(t)^{\frac{N}{2}} u(t, \tilde{\lambda}(t)(\cdot - \tilde{\gamma}(t)))e^{i\theta(t)} - Q(\cdot) \right\| \to 0 \quad \text{as} \quad t \to T_m \tag{2.1}$$

for some $\tilde{y}(t) \in \mathbb{R}^N$ and $\theta(t) \in \mathbb{R}$ (Weinstein [37]).

(II) Let $u(t)$ be a solution of (NSC) such that $xu(t) \in L^2(\mathbb{R}^N)$ and $\|\nabla u(t)\| \to \infty$ as $t \to T_m$ for some $T_m \in (0, \infty)$. If $u(t)$ satisfies $\|(x-a)u(t)\| \to 0$ $(t \to T_m)$, then $u(t)$ must be of the form:

$$(T_m - t)^{-N/2} \exp\left(\frac{-i|x_t(a,v)|^2}{2(T_m - t)}\right) V\left(\frac{t}{T_m(T_m - t)}, \frac{x_t(a,v)}{T_m - t}\right) e^{\frac{iv}{T_m}(x - \frac{v}{2T_m}t)}, \tag{2.2}$$

where $V(t,x)$ is also a solution of (NSC) in $C(\mathbb{R}_+; H^1(\mathbb{R}^N) \cap L^2(\mathbb{R}^N; |x|^2 dx))$ such that $E(V(t)) = 0$, and where

$$x_t(a,v) = x - a + v - \frac{v}{T_m}t$$

for an appropriate $v \in \mathbb{R}^N$ (Nawa and M.Tsutsumi [29]).

(III) For given L points $\{a^1, a^2, \cdots, a^L\} \subset \mathbb{R}^N$, there exist a blow-up solution $u(t)$ of (NSC) such that

$$\left\| u(t) - \sum_{j=1}^{L} Q^j(t) \right\|_\sigma \to 0 \quad \text{as} \quad t \to T_m, \tag{2.3}$$

where

$$Q^j(t,x) = (T_m - t)^{-N/2} \exp\left(\frac{-i|x - a^j|^2}{2(T_m - t)}\right) Q\left(\frac{x - a^j}{T_m - t}\right) e^{it/2T_m(T_m - t)} \tag{2.4}$$

for $T_m \in (0, \infty)$ (Merle [20]).

Remark 2.1. $Q(x)e^{it/2}$ which is a standing wave solution of (NSC), is transformed into Q^j by the space-time transformation appearing in the left hand side of (2.2) with $a = a^j$ and $v = 0$. We call this transformation pseudo-conformal transformation. Since we have $E(Q(\cdot)e^{it/2}) = 0$, Q^j is a blow-up solution of (NSC) such that $\|(x - a^j)u(t)\| \to 0$ $(t \to T_m)$ by virtue of (II).

These results require additional conditions on initial data (or solutions): $\|u_0\| = \|Q\|$ for (I); $|x|u_0 \in L^2(\mathbb{R}^N)$ for (II) and (III). Our purpose here is to investigate the asymptotic profile of generic H^1 blow-up soluton of (NSC).

Now we state our main result [27,28]:

Theorem B. *Let $u(t)$ be the singular solution of (NSC) such that*

(B.1) $$\limsup_{t \to T_m} \|\nabla u(t)\| = \limsup_{t \to T_m} \|u(t)\|_\sigma = \infty$$

for some $T_m \in (0, \infty]$. Let $\{t_n\}$ be any sequence such that

(B.2) $$\sup_{t \in [0, t_n]} \|u(t)\|_\sigma = \|u(t_n)\|_\sigma.$$

For this $\{t_n\}$, we put

(B.3) $$\lambda_n = \frac{1}{\|u(t_n)\|_\sigma^{\sigma/2}}$$

and, we consider the scaled functions

(B.4) $$u_n(t,x) = \lambda_n^{\frac{N}{2}} \overline{u(t_n - \lambda_n^2 t, \lambda_n x)}$$

for $t \in [0, t_n/\lambda_n^2)$. Then there exists a subsequence of $\{u_n\}$ (still denoted by $\{u_n\}$), which satisfies the following properties: there exist

(i) *a finite number of nontrivial solutions $u^1, u^2, \cdots, u^L$ of (NSC) in $C_b(\mathbb{R}_+; H^1(\mathbb{R}^N))$ with $E(u^j) = 0$, and*

(ii) *sequences $\{\gamma_n^1\}, \{\gamma_n^2\}, \cdots, \{\gamma_n^L\}$ in $\mathbb{R}^N$,*

such that, for any $T > 0$,

$$\lim_{n\to\infty} \sup_{t\in[0,T]} \left\| u_n(t,\cdot) - \sum_{j=1}^{L} u^j(t, \cdot - \gamma_n^j) \right\|_\sigma = 0, \tag{B.5}$$

$$\lim_{n\to\infty} \sup_{t\in[0,T]} \left\| \nabla u_n(t,\cdot) - \sum_{j=1}^{L} \nabla u^j(t, \cdot - \gamma_n^j) \right\| = 0, \tag{B.6}$$

$$\lim_{n\to\infty} \sup_{t\in[0,T]} \left\| u_n(t,\cdot) - \sum_{j=1}^{L} u^j(t, \cdot - \gamma_n^j) - \phi_n(t,\cdot) \right\| = 0, \tag{B.7}$$

where

$$\phi_n(t,\cdot) = \exp(\frac{it}{2}\Delta) * \left(u_n(0,\cdot) - \sum_{j=1}^{L} u^j(0, \cdot - \gamma_n^j) \right). \tag{B.8}$$

Furthermore we have

$$\|u_0\|^2 \geqq \sum_{j=1}^{L} \|u^j(t)\|^2 \geqq L\|Q\|^2. \tag{B.9}$$

where Q is a nontrivial solution of (A.1) and (A.2).

Remark 2.2. (1) If the solution satisfies $\limsup_{t\to T_m} \|\nabla u(t)\| = \infty$, then we have, by the energy conservation law $\limsup_{t\to T_m} \|u(t)\|_\sigma = \infty$. So, (B.1) is always assured. If $T_m < \infty$, we have (B.1) with limsup replaced by lim.

(2) We can choose a sequence as in (B.2), since we have (B.1).

(3) The scaled function u_n in (B.4) also solves (NSC), and satisfies $\|u_n(t)\| = \|u(t)\|$ and $E(u_n(t)) = \lambda_n^2 E(u(t))$. This is a special feature of (NSC).

Each u^j can be considered to correspond to the "strong" singularity in blow-up solution, since one has, by (B.7),

$$\lim_{n\to\infty} \sup_{t\in[t_n-\lambda_n^2 T, t_n]} \left\| \overline{u(t,\cdot)} - \sum_{j=1}^{L} u_n^j(t,\cdot) - \tilde{\phi}_n(t,\cdot) \right\| = 0, \tag{2.5}$$

where

$$u_n^j(t,x) = \frac{1}{\lambda_n^{N/2}} u^j \left(\frac{t_n - t}{\lambda_n^2}, \frac{x - \gamma_n^j \lambda_n}{\lambda_n} \right), \tag{2.6}$$

$$\tilde{\phi}_n(t,x) = \frac{1}{\lambda_n^{N/2}} \phi_n \left(\frac{t_n - t}{\lambda_n^2}, \frac{x}{\lambda_n} \right). \tag{2.7}$$

We note that there is a possibility that $\tilde{\phi}_n$ produce "weak" singularities, around which the rate of blow-up is lower than $\|u(t)\|_\sigma$. Thus we may say that the term $\sum_{j=1}^{L} u_n^j(t,\cdot)$ is the top term of the asymptotic expansion of blow-up solution near the blow-up time T_m. If $N \geqq 2$, "weak" singularities may form a $N-1$ dimensional manifold as in the case of semilinear heat equations (Giga and Kohn [10]). If $\tilde{\phi}_n$ produce no singularity, we can safely say that the blow-up set consists of finite number of points as in the case of one-dimensional semilinear heat equations (Chen and Matano [8]).

However, there still remains a possibility of "ergodic" or "chotic" behavior of singularities, since (as is mentioned in the ending of the previous section) the solution of nonlinear Schrödinger equation involves a ***many body*** character in its behavior. Each "singularity" $\{u_n^j\}$ considered to be "a cluster" or "a particle" in the analogy of many body Schrödinger equation. If the remainder term $\tilde{\phi}_n(t,x)$ plays a role of long range potentials, the motion of "clusters" will be very complicated. Thus the expression of Theorem B using the sequence of scaled solutions (B.4) may be a considerable one.

Remark 2.3. As in the proof of [Theroem C;24], we can show that if $u(0) \in H^1(\mathbb{R}^N) \cap L^2(|x|^2dx)$, then

$$\sup_{n\in\mathbb{N}} |\gamma_n^j \lambda_n| < \infty \quad (j = 1, 2, \cdots, L) \tag{1.10}$$

in Theorem B.

3. SKETCH OF PROOF OF THEOREM B

Theorem B seems to be closely related to a phenomenon which has been observed in various nonlinear problems by the name of bubble theorem or concentrated compactness theorem (for example, see [4,16,17,18,22,33,34]). In fact, the proof of this theorem is inspired by Brézis and Coron [4]. One may find that the underlying idea being the method of concentrated compactness due to Lions [17,18]. (The use of the general method of concentrated compactness in the study of blow-up problem for the nonlinear Schrödinger equation can be traced back to Weinstein [37].) However, we do not use the general method of it. Our basic tool is the compactness device as in Lieb [16] (see also Brézis and Lieb [6] and Fröhlich, Lieb and Loss [9]). We extend Lieb's compactness lemma to space-time one, with which the Ascoli-Arzela theorem plays a crucial role in our analysis working with the scaled solutions of (NSC) defined by (B.4). The following proposition is the heart of the matter.

Proposition 3.1. *Let $\{v_n\}$ be an equibounded family in $C([0,T]; H^1(\mathbb{R}^N))$ such that*

$$2i\frac{\partial v_n}{\partial t} + \Delta v_n + |v_n|^{\frac{4}{N}} v_n = g_n, \tag{3.1}$$

$$\sup_{t\in[0,T]} \|v_n(t)\|_\sigma \neq 0. \tag{3.2}$$

Here $\{g_n\}$ is an equibounded family in $C([0,T]; L^{\sigma'}(\mathbb{R}^N))$ such that, for any $R > 0$,

$$\lim_{n\to\infty} \sup_{t\in[0,T]} \|g_n(t,\cdot)\|_{\sigma'} = 0, \tag{3.3}$$

where $\frac{1}{\sigma} + \frac{1}{\sigma'} = 1$. Then there exists

(i) *a nontrivial solution v of (NSC) in $C([0,T]; H^1(\mathbb{R}^N))$ and*

(ii) *a sequence $\{\gamma_n\} \subset \mathbb{R}^N$*

such that for $\Omega \Subset \mathbb{R}^N$ and for some subsequence (still denoted by the same letter),

$$\tilde{v}_n \equiv v_n(\cdot, \cdot + \gamma_n) \overset{*}{\rightharpoonup} v \quad \text{weakly* in } L^\infty([0,T]; H^1(\mathbb{R}^N)), \tag{3.4}$$

$$\tilde{v}_n \to v \quad \text{strongly in } C([0,T]; L^\alpha(\Omega)) \text{ for } \alpha \in [2, 2^*) \tag{3.5}$$

as $n \to \infty$.

Furthermore we have

$$|\tilde{v}_n|^{\frac{4}{N}}\tilde{v}_n - |\tilde{v}_n - v|^{\frac{4}{N}}(\tilde{v}_n - v) - |v|^{\frac{4}{N}}v \to 0 \quad \text{strongly in } C([0,T]; L^{\sigma'}(\mathbb{R}^N)), \tag{3.6}$$

$$\lim_{n\to\infty} \sup_{t\in[0,T]} \int_{\mathbb{R}^N} \big||\tilde{v}_n|^\alpha - |\tilde{v}_n - v|^\alpha - |v|^\alpha\big| dx = 0, \quad \alpha \in [2, 2^*), \tag{3.7}$$

$$\lim_{n\to\infty} \int_0^T \{E(\tilde{v}_n) - E(\tilde{v}_n - v) - E(v)\}\, dt = 0, \tag{3.8}$$

and for any $t \in [0,T]$

$$\lim_{n\to\infty} \{E(\tilde{v}_n(t)) - E((\tilde{v}_n - v)(t)) - E(v(t))\} = 0. \tag{3.9}$$

We can safely say that our analysis investigates, by means of Proposition 3.1, how the "dichotomy" (in the terminology of concentrated compactness) occurs in the sequence $\{u_n\}$. Theorem B asserts that u_n behaves like a finite superposition of *zero energy* time global solutions of (NSC) (see (B.5) - (B.7)).

One can easily see that the scaled function u_n satisfies

$$2i\frac{\partial u_n}{\partial t} + \Delta u_n + |u_n|^{\frac{4}{N}}u_n = 0, \tag{3.10}$$

$$\|u_n(t)\| = \|u_0\|, \tag{3.11}$$

$$\sup_{t\in[0,T]} \|u_n(t)\|_\sigma = 1 \quad \text{for any} \quad T > 0, \tag{3.12}$$

$$E(u_n(t)) = \lambda_n^2 E(u_0)\,(\to 0) \quad \text{as } n \to \infty. \tag{3.13}$$

From (3.10), (3.11) and (3.12), it follows that $\{u_n\}$ is an equi-bounded family in $L^\infty(\mathbb{R}_+; H^1(\mathbb{R}^N))$. Thus, by Proposition 3.1, there exist

(i) a nontrivial solution u^1 of (NSC) in $C([0,\infty); H^1(\mathbb{R}^N))$

(ii) a sequence $\{y_n^1\} \subset \mathbb{R}^N$

such that for $\Omega \Subset \mathbb{R}^N$ and for some subsequence (still denoted by the same letter),

$$u_n^1 \equiv u_n(\cdot, \cdot + y_n^1) \overset{*}{\rightharpoonup} u^1 \text{ weakly* in } L^\infty([0,\infty); H^1(\mathbb{R}^N)), \tag{3.14}$$

$$u_n^1 \to u^1 \quad \text{strongly in } C([0,T]; L^\alpha(\Omega)) \tag{3.15}$$

for $\alpha \in [2, 2^*)$ as $n \to \infty$. Furthermore we have

$$|u_n^1|^{\frac{4}{N}}u_n^1 - |u_n^1 - u^1|^{\frac{4}{N}}(u_n^1 - u^1) - |u^1|^{\frac{4}{N}}u^1 \to 0 \tag{3.16}$$

strongly in $C([0,T];L^{\sigma'}(\mathbb{R}^N))$,

$$\lim_{n\to\infty}\sup_{t\in[0,T]}\int_{\mathbb{R}^N}\left||u_n^1|^\alpha-|u_n^1-u^1|^\alpha-|u^1|^\alpha\right|dx=0,\quad \alpha\in[2,2^*), \tag{3.17}$$

$$\lim_{n\to\infty}\int_0^T\{E(u_n^1)-E(u_n^1-u^1)-E(u^1)\}\,dt=0, \tag{3.18}$$

and, for any $t\in\mathbb{R}_+$

$$\lim_{n\to\infty}\{E(u_n^1(t))-E((u_n^1-u^1)(t))-E(u^1(t))\}=0. \tag{3.19}$$

Suppose that $\limsup_{n\to\infty}\|u_n^1-u^1\|_\sigma\neq 0$. $u_n^1-u^1$ satisfies

$$2i\frac{\partial(u_n^1-u^1)}{\partial t}+\Delta(u_n^1-u^1)+|u_n^1-u^1|^{\frac{4}{N}}(u_n^1-u^1)=g_n^1, \tag{3.20}$$

where

$$g_n^1(t,x)=-(|u_n^1|^\sigma-|u_n^1-u^1|^\sigma-|u^1|^\sigma)(t,x). \tag{3.21}$$

We note here that (3.16) implies that for any $\{x_n\}\subset\mathbb{R}^N$ and for any $T>0$,

$$\lim_{n\to\infty}\sup_{t\in[0,T]}\|g_n^1(t,\cdot+x_n)\|_\sigma=0. \tag{3.22}$$

Clearly $\{u_n^1-u^1\}$ is an equi-bounded family in $L^\infty([0,\infty);H^1(\mathbb{R}^N))$. At this stage we apply Proposition 3.1 to $\{u_n^1-u^1\}$.

We iteratively use Proposition 3.1 to construct u^j's, and we have

$$\lim_{n\to\infty}\{E(u_n(t))-E((u_n^j-u^j)(t))\}=\sum_{k=1}^{j}E(u^k(t)). \tag{3.23}$$

Here the important thing is the finiteness of u^j's. This follows from $E(u^j)=0$ (forany j), since in this case we have $\|u^j\|\geqq\|Q\|$ for any j (see the definition of Q in Theorem A). If the iteration were not terminated at some finite index, we would have by the construction of u^j's that

$$\lim_{j\to\infty}\lim_{n\to\infty}\sup_{t\in[0,T]}\|(u_n^j-u^j)(t)\|_\sigma=0. \tag{3.24}$$

The formula (3.23) together with (3.13) and (3.24) yields that

$$\lim_{j\to\infty}\sum_{k=1}^{j}E(u^k(t))\leqq\lim_{j\to\infty}\lim_{n\to\infty}\sup_{t\in[0,T]}\|(u_n^j-u^j)(t)\|_\sigma=0. \tag{3.25}$$

Hence we can conclude $E(u^j)=0$ for any j, if we know that every bounded global solution must have nonnegative energy, *i.e.*;

Theorem C. *If the solution $u(t)$ of (NSC) belongs to $C_b([0,\infty);H^1(\mathbb{R}^N))$, then its energy must be nonnegative, i.e.,*

$$E(u(0))\geqq 0.$$

In other words, if $E(u(0)) < 0$, then there exists $T_m \in (0, \infty]$ such that the corresponding solution $u(t)$ of (NSC) satisfies

$$\limsup_{t \to T_m} \|\nabla u(t)\| = \limsup_{t \to T_m} \|u(t)\|_\sigma = \infty,$$

i.e., $u(t)$ blows up (in finite time) or grows up (at infinity). If $T_m < \infty$, we can replace lim sup *by* lim.

This is an improvement of previous results concerning the existence of blow-up solution of (NSC) in the sence that we do not require additional conditions on initial data except $E(u(0)) < 0$. However Theorem C does not assert that every negative energy initial datum leads to the blow-up solution of $C(1 + \frac{4}{N})$. There remains a possibility that $T_m = \infty$. The blow-up of negative energy solutions has been proved under some conditions: $|x|u_0 \in L^2(\mathbb{R}^N)$ (Glassey [13]) while this is an important class of initial data and quite reasonable physically; $N \geqq 2$ and u_0 is radially symmetric (Ogawa-Y.Tsutsumi[30]); $N = 1$ and $p = 1 + \frac{4}{N}$ (Ogawa-Y.Tsutsumi[31]). Hence the results of Ogawa-Tsutsumi [30,31] ensure that Theorem C with $T_m < \infty$ holds true for the case of $N = 1$, and of $N \geqq 2$ and $u(0)$ being radially symmetric. We refer [28] for the proof of Theorem B ($N \geqq 2$), although we do not mention the proof of it in this paper. We may say that Theorem C is a weak version of a theorem which has long been speculated.

ACKNOWLEDGEMENT. The author thanks professors Drumi Bainov and Joji Kajiwara for their kind invitation to participate in the 2nd International Colloquium on Numerical Analysis and 4th International Colloquim on Differential Equations. The travel and living expenses of the author were partially supported by Multi-disciplinary Science Foundation.

References

1. S.A. Akhmanov, A.P.Sukhorukov and R.V. Khokhlov, *Self-focusing and self-trapping of intence light beams in a nonlinear medium*, Soviet physics JETP **23** (1966), 1025–1033.
2. ———, *Self-focusing and diffraction of light in a nonlinear medium*, Soviet physics USPEKHI **93** (1968), 609–633.
3. H. Berestycki and P.L. Lions, *Nonlinear scalar field equations*. I, II, Archi. Rat. Mech. Anal. **82** (1983), 313–376.
4. H. Brézis and J.M. Coron, *Convergence of H−system or how to blow bubbles*, Archi. Rat. Mech. Anal. **89** (1985), 21–56.
5. H. Brézis and E.H. Lieb, *A relation between pointwise convergence of functions and convergence of functionals*, Proc. Am. Math. Soc. **88** (1983), 485–489.
6. ———, *Minimum action solutions of some vector field equations*, Commun. Math. Phys. **96** (1983), 97–113.
7. T. Cazenave and F.B. Weissler, *The structure of solution to the pseudoconformaly invariaint nonlinear Schrödinger equations*, preprint.
8. X.-Y.Chen and H. Matano, *Convergence, asymptotic periodicity, and finite-point blow-up in one-dimensional semilinear heat equations*, J. Differential Equations **104** (1986), 251–270.
9. J. Fröhlich, E.H. Lieb and M. Loss, *Stability of coulomb systems with magnetic fields I, The one- electron atom*, Commun. Math. Phys. **78** (1989), 160–190.
10. Y. Giga and R.V. Kohn, *Nondegeneracy of blowup for semilinear heat equations*, Commun. Pure Appl. Math. **17** (1989), 845–884.
11. J. Ginibre and G. Velo, *On a class of nonlinear Schrödinger equations*. I, II, J. Funct. Anal. **32** (1979), 1–71.
12. ———, *The global Cauchy problem for the nonlinear Schrödinger equation revisited*, Ann. Inst. Henri Poincaré, Analyse Nonlinéaire. **2** (1985), 309–327.
13. R.T. Glassey, *On the blowing up solution to the Cauchy problem for nonlinear Schrödinger equations*, J. Math. Phys. **18** (1979), 1794–1797.
14. T. Kato, *On nonlinear Schrödinger equations*, Ann. Inst. Henri Poincaré, Physique Theorique **46** (1987), 113–129.
15. B. LeMesurier, C. Papanicolau, C. Sulem and P.L. Sulem, *The focusing singularity of the nonlinear Schrödinger equation*, Direction in Partial Differential Equations (Crandall,M.G., Rabinowitz, P.H. and Turner,R.E., eds.), Academic Press, New York, 1987, pp. 159–201.
16. E.H. Lieb, *On the lowest eigenvalue of Laplacian for the intersection of two domains*, Invent. Math. **74** (1983), 441–448.
17. P.L. Lions, *The concentration compactness principle in the calculus of variations. The locally compact case*. I, II, Ann. Inst. Henri Poincaré, Analyse Nonlinéaire. **1** (1984), 109–145, 223–283.
18. ———, *The concentration compactness principle in the calculus of variations. The limit case*. I, II., Riv. Math. Iberoamericana **1** (1985), 45–121, 145–201.

19. F. Merle, *Limit of the solution of the nonlinear Schrödinger equation at the blow-up time*, J. Funct. Anal. **84** (1989), 201–214.
20. ______, *Construction of solutions with exactly k blow-up points for the Schrödinger equation with the critical power nonlinearity*, Commun. Math. Phys. **129** (1990), 223–240.
21. F. Merle and Y. Tsutsumi, *L^2 concentration of blow-up solutions for the nonlinear Schödinger equation with the critical power nonlinearity*, J. Differential Equations **84** (1990), 205–214.
22. K. Nagasaki and T. Suzuki, *Asymptotic analysis for two-dimensional elliptic eigenvalue problems with exponentially dominated nonlinearities*, Asymptotic Analysis **3** (1990), 173–188.
23. H. Nawa, *"Mass concentration" phenomenon for the nonlinear Schrödinger equation with the critical power nonlinearity*, Funk. Ekva. **35** (1992), 1–18.
24. ______, *"Mass concentration" phenomenon for the nonlinear Schrödinger equation with the critical power nonlinearity. II*, Kodai Math. J. **13** (1990), 333–348.
25. ______, *Convergence theorems for the pseodo-conformally invariant nonlinear Schrödinger equation*, Proc. Japan Acad. **66 (A)** (1990), 214–216; RIMS Kokyuroku **755** (1991), 73–92.
26. ______, *Formation of singularities in solutions of the nonlinaer Schrödinger equation*, Proc. Japan Acad. **67 (A)** (1991), 29–34.
27. ______, *Asymptotic profiles of blow-up solutions of the nonlinear Schrödinger equation with critical power nonlinearity*, J. Math. Soc. Japan (to appear).
28. ______, *Asymptotic profiles of blow-up solutions of the nonlinear Schrödinger equation*, Singularities in fluids, plasmas and optics (R. E. Caflisch and G. C. Papanicolau, eds.), NATO ASI series, Kluwer, Dordrecht, 1993, pp. 221–253; its revised manuscript, T.I.T. preprint series, 1993 (to appear).
29. H. Nawa and M. Tsutsumi, *On blow-up for the pseudo-conformally invariant nonlinear Schrödinger equation*, Funk. Ekva. **32** (1989), 417–428.
30. T .Ogawa and Y. Tsutsumi, *Blow-up of H^1-solution for the nonlinear Schrödinger equation*, J. Differential Equations **92** (1991), 317–330.
31. ______, *Blow-up of H^1-solution for the one dimensional nonlinear Schrödinger equation with critical power nonlinearity*, Proc. Amer. Math. Soc. **111** (1991), 487–496.
32. W. A. Strauss, *Existence of solitary waves in higher dimensions*, Commun. Math. Phys. **55** (1977), 149–162.
33. M. Struwe, *A global compactness result for elliptic boundary value problem involving limiting nonlinearities*, Math. Z. **187** (1983), 567–576.
34. S. Takakuwa, *Behavior of minimizing sequences for the Yamabe problem*, Tokyo Metropolitan University Mathematics Preprint Series **7** (1990).
35. Y. Tsutsumi, *Rate of L^2 concentration of blow-up solutions for the nonlinear Schödinger equation with critical power nonlinearity*, Nonlinear Anal. T.M.A. **15** (1990), 719–724.
36. M.I. Weinstein, *Nonlinear Schrödinger equations and sharp interpolation estimates.*, Commun. Math. Phys. **87** (1983), 511–517.
37. ______, *On the structure and formation singularities in solutions to nonlinear dispersive evolution equations*, Commun. in Partial Diffrential Equations **11** (1986), 545–565.
38. ______, *The nonlinear Schrödinger equation - Singularity formation, Stability and dispersion*, The Connection between Infinite and Finite Dimensional Dynamical Systems, Contemporary Math.99, 1989, pp. 213–232.
39. K. Yajima, *Existence of solutions for Schrödinger evolution equations*, Commun. Math. Phys. **110** (1987), 415–426.
40. V.E. Zakharov and V.S. Synakh, *The nature of self-focusing singularity*, Sov. Phys. JETP **41** (1976), 441–448.

2nd Int. Coll. on Numerical Analysis, pp. 143-153
D. Bainov and V. Covachev (Eds)

Approximation by Univariate and Bivariate Splines

G. Nürnberger

Fakultät für Mathematik und Informatik,
Universität Mannheim, 68131 Mannheim, Germany

Abstract. A short survey of best uniform approximation by splines with free knots and of Lagrange and Hermite interpolation by bivariate spline spaces is given. Some typical results are discussed and the basic ideas of certain numerical methods are described. In contrast to splines with fixed knots, some difficult problems are unsolved at present, in these fields.

Keywords. Splines with Fixed Knots, Splines with Free Knots, Bivariate Splines, Best Approximation, Interpolation.

Introduction

There is a fully developed theory of interpolation and best approximation by splines with fixed knots including efficient numerical methods (see the books of de Boor [4], Schumaker [64] and Nürnberger [37]). On the other hand, some difficult problems are unsolved at present if analogous questions are considered for splines with free knots and for bivariate splines.

The reason is that best uniform approximation by splines with free knots is a highly non-linear problem. In contrast to rational functions (see the book of Braess [9]), local and global best spline approximations with free knots are not the same, in general. Therefore, if Newton type methods are applied, then the convergence is not guaranteed and only local best approximations are computed, in general. However, an algorithm for computing efficient spline approximations with free knots which converges by starting with an arbitrary set of knots is available.

On the other hand, the theory of free knot splines is not fully developed. In particular, a characterization of best spline approximations is not known. Alternation characterizations were given for the subclass of strongly unique best approximations. Moreover, results are known on the semi-continuity of the set-valued metric projection, but, for example, the existence of continuous selections is unsolved at present. We discuss some typical results which we compare with the fixed knot case.

The reason why there is no complete theory of approximation by bivariate spline spaces analogous to the univariate case is that these spaces possess a more complex structure. Even basic problems such as the dimension and basis of bivariate spline spaces on arbitrary triangulations are not completely solved. However, the dimension and basis is known for cross-cut partitions (e.g. uniform partitions) of the domain, and in the case of non-uniform triangulations for splines of certain degree and smoothness (see the survey of Schumaker [66] and the book of Chui [11]).

In this paper, we concentrate on Lagrange and Hermite interpolation by bivariate spline spaces. Several authors investigated this problem in the case of splines of low degree using Bernstein-Bézier techniques. Recently, a different method was developed for constructing interpolation points for spaces of splines of abitrary degree and smoothness. The basic principle is that the bivariate problem is reduced to certain univariate problems. In standard cases, the interpolating spline can be computed by passing from one triangle to the next solving only small systems, and this spline yields (nearly) optimal approximation order.

For the study of quasi-interpolation and cardinal interpolation, the reader is referred to the book of Chui [11] and the references therein.

1 Best Approximation by Splines with Free Knots

Basic questions in the theory of best approximation are existence, characterization, uniqueness, computation and approximation order of best approximations. Concerning the approximation order of splines, the reader is referred to the book of Schumaker [64] and the surveys of DeVore [19] and Popov [54].

We begin by recalling some typical complete results on splines with fixed knots in order to compare them later with the free knot case. Let knots $a = x_0 < x_1 < \cdots < x_k < x_{k+1} = b$ be given. We call $S_m(x_1, \ldots, x_k) = \{s \in C^{m-1}[a,b] : s|_{[x_i,x_{i+1}]} \in \Pi_m,\ i = 0, \ldots, k\}$ the space of *splines of degree* m *with* k *fixed knots* $x_1, \ldots, x_k$. Here Π_m denotes the space of *polynomials of degree at most* m.

First, we consider Lagrange interpolation by splines from $S_m(x_1, \ldots, x_k)$. The *Lagrange interpolation problem* is to determine for $f \in C[a,b]$ and $a \leq t_1 < \cdots < t_{m+k+1} \leq b$, a spline $s \in S_m(x_1, \ldots, x_k)$ such that $s(t_i) = f(t_i)$, $i = 1, \ldots, m+k+1$. (Note, that the dimension of $S_m(x_1, \ldots, x_k)$ is $m+k+1$.)

The first theorem due to Schoenberg & Whitney [60] characterizes those sets of points which admit unique Lagrange interpolation by a simple interlacing condition. An analogous result holds for Hermite interpolation (see e.g. [4], [64], [37]).

Theorem 1.1. *The following statements are equivalent:*

(i) *For all* $f \in C[a,b]$, *the above Lagrange interpolation problem has a unique solution from* $S_m(x_1, \ldots, x_k)$.

(ii) $t_j < x_j < t_{j+m+1}$, $j = 1, \ldots, k$.

The Schoenberg-Whitney theorem plays an important role in the study of best approximations. The *problem of best approximation* is to determine for $f \in C[a,b]$, a spline $s_f \in S_m(x_1, \ldots, x_k)$ such that $\|f - s_f\| = \inf_{s \in S_m(x_1,\ldots,x_k)} \|f - s\|$, where $\|h\| = \sup_{t \in [a,b]} |h(t)|$ denotes the uniform norm. This spline s_f is called *best approximation* of f from $S_m(x_1, \ldots, x_k)$, and $d(f, S_m(x_1, \ldots, x_k)) = \inf_{s \in S_m(x_1,\ldots,x_k)} \|f - s\|$ is called the *minimal deviation* of f from $S_m(x_1, \ldots, x_k)$.

It is well known that best approximations from $S_m(x_1, \ldots, x_k)$ always exist, since the space has finite dimension. Best approximations are characterized by alternation properties of the error function. Given $h \in C[a,b]$, we call points $a \leq t_1 < \cdots < t_p \leq b$ *alternating extreme points* of h if there exists a sign $\sigma \in \{-1, 1\}$ such that $\sigma(-1)^i h(t_i) = \|h\|$, $i = 1, \ldots, p$. If I is a subinterval of $[a,b]$, then we denote by $A(h)|_I$ the maximal number of alternating extreme points of h on I.

In the following, let $f \in C[a,b]$ and $s_f \in S_m(x_1, \ldots, x_k)$ be given. We begin with a characterization of best approximations which was proved independently by Rice [56] and Schumaker [61].

Theorem 1.2. *The following statements are equivalent:*

(i) *The spline* s_f *is a best approximation of* f *from* $S_m(x_1, \ldots, x_k)$.

(ii) *An interval* $[x_p, x_{p+q+1}] \subset [a,b]$, $q \geq 0$, *exists such that* $A(f - s_f)|_{[x_p,x_{p+q+1}]} \geq m + q + 2$.

Moreover, it was shown by Berens & Nürnberger [3] that for each $f \in C[a,b]$, the non-empty set of points on which all best approximations of f from $S_m(x_1,\ldots,x_k)$ coincide is the union of knot-intervals. In general, best spline approximations are not unique.

We first consider the strong unicity of best approximations. A spline $s_f \in S_m(x_1,\ldots,x_k)$ is called *strongly unique best approximation* of $f \in C[a,b]$ if there exists a constant $K_f > 0$ such that for all $s \in S_m(x_1,\ldots,x_k)$, $\|f-s\| \geq \|f-s_f\| + K_f\|s-s_f\|$. The next result due to Nürnberger [34] shows that strongly unique best approximations can also be characterized by alternation properties of the error function. In order to state the result, we consider arbitrary additional knots $x_{-m} < \cdots < x_{-1} < a$ and $b < x_{k+2} < \cdots < x_{k+m+1}$.

Theorem 1.3. *The following statements are equivalent:*

(i) *The spline s_f is a strongly unique best approximation of f from $S_m(x_1,\ldots,x_k)$.*

(ii) *For all $(x_i, x_{i+m+j}) \subset (x_{-m}, x_{k+m+1})$, $j \geq 1$, we have $A(f-s_f)|_{(x_i,x_{i+m+j})} \geq j+1$.*

Building on earlier work of Rice [56], Schumaker [61] and Strauss [72], a characterization of unique best approximations from $S_m(x_1,\ldots,x_k)$ was given by Nürnberger & Singer [47] which is somewhat different to Theorem 1.3.. An iterative method for computing best spline approximations from $S_m(x_1,\ldots,x_k)$ was developed by Nürnberger & Sommer [49]. As in the classical Remez algorithm for Π_m, in the n-th step a spline $s_n \in S_m(x_1,\ldots,x_k)$ and a real number λ_n has to be computed such that $(-1)^i\,(f(t_{i,n}) - s_n(t_{i,n})) = \lambda_n$, $i = 1,\ldots,m+k+2$. Then a point in $M_n = \{t_{1,n},\ldots,t_{m+k+2,n}\}$ is replaced by an extremal point of $f - s_n$ such that a new set M_{n+1} is obtained. The basic problem is that, in contrast to Π_m, the new system may not be solvable. In [10], an appropriate exchange rule was described such that the new system has a unique solution. Moreover, it was proved that, roughly speaking, the sequence (s_n) converges to a spline $s_f \in S_m(x_1,\ldots,x_k)$ such that $(-1)^i\,(f(t_i) - s_f(t_i)) = \|f-s_f\|$, $i = 1,\ldots,m+k+2$. Then it follows from Theorem 1.2. that s_f is a best approximation of f from $S_m(x_1,\ldots,x_k)$ (see also [37]).

The above results were extended by Nürnberger, Schumaker, Sommer & Strauss [43], [44], [45] and Freyburger [20] to certain spaces of generalized splines, where the space Π_m is replaced by Haar spaces.

This briefly indicates that there is a fully developed theory of approximation by splines with fixed knots. For further results, the reader is referred to Berens & Nürnberger [3], Pinkus & Strauss [53], Sommer & Strauss [70], [71] and to Nürnberger [37]. In the following, we consider the free knot case for which the situation is different.

We denote by $S_{m,k}$ the set of *splines of degree m with k free knots*. Here, the k knots of the splines can be chosen arbitrarily in the considered interval $[a,b]$. But in order to guarantee the existence of best approximations, splines with multiple knots have to be included in the set $S_{m,k}$. (For a precise definition of $S_{m,k}$ see e.g. [37].) Since $S_{m,k}$ is a non-convex set, best approximation by $S_{m,k}$ is a non-linear problem.

In the following, we assume for brevity that $f \in C[a,b]$, $m \geq 2$ and $s_f \in S_m(x_1,\ldots,x_k)\backslash S_{m,k-1}$, i.e. s_f has k simple knots. (We note that the set $S^1_{m,k}$ of splines of degree m with k free simple knots is an open and dense subset of $S_{m,k} \cap C[a,b]$.)

The existence of continuous best approximations from $S_{m,k}$ for continuous functions was established by Schumaker [62] (see also [63], Rice [57]). One of the main open problems is the characterization of best approximations from $S_{m,k}$. We will discuss the following result (cf. Theorem 1.2.).

Theorem 1.4. *We consider the following statements:*

(i) *An interval $[x_p, x_{p+q+1}] \subset [a,b]$, $q \geq 0$, exists such that $A(f-s_f)|_{[x_p,x_{p+q+1}]} \geq m+k+q+2$.*

(ii) *The spline s_f is a best approximation of f from $S_{m,k}$.*

(iii) *An interval $[x_p, x_{p+q+1}] \subset [a,b]$, $q \geq 0$, exists such that $A(f - s_f)|_{[x_p,x_{p+q+1}]} \geq m + 2q + 2$.*

Then (i) $\underset{\not\Leftarrow}{\Rightarrow}$ (ii) $\underset{\not\Leftarrow}{\Rightarrow}$ (iii).

Building on an earlier result of Schumaker [62], the implication (i) ⇒ (ii) was proved by Braess [8] for splines with multiple knots. By using the tangent method of G. Meinardus and D. Schwedt (see Meinardus [30]), (ii) ⇒ (iii) was established by Braess [8] for differentiable splines. The implication (ii) ⇒ (iii) was shown by Nürnberger, Schumaker, Sommer & Strauss [46] for non-differentiable splines. Recently, a stronger necessary condition for best approximations from $S_{m,k}$ was developed by Mulansky [32], [33] by using an appropriate tangent cone instead of the classical tangent space. By applying optimization techniques, Kawasaki [27] investigated the connection of (iii) to nearly best local approximations from $S_{m,1}$.

On the other hand, a complete alternation characterization of the subclass of strongly unique best approximations with simple knots was given by Nürnberger [36] (cf. Theorem 1.3.). In Theorem 1.5., we denote by $y_1 \leq \cdots \leq y_{2k}$ the knots $x_1 < \cdots < x_k$ of s_f, counting each knot twice, and choose additional knots $y_{-m} < \cdots < y_{-1} < y_0 = a$ and $b = y_{2k+1} < y_{2k+2} < \cdots < y_{2k+m+1}$.

Theorem 1.5. *The following statements are equivalent:*

(i) *The spline s_f is a strongly unique best approximation of f from $S_{m,k}$.*

(ii) *For all $(y_i, y_{i+m+j}) \subset (y_{-m}, y_{2k+m+1})$, $j \geq 1$, we have $A(f - s_f)|_{(y_i,y_{i+m+j})} \geq j + 1$.*

Moreover, it was shown by Nürnberger [39] that the (strong) unicity condition of Schumaker [62], Arndt [2] and Schaback [59], which says $A(f-s_f)|_{[a,b]} \geq m+2k+2$, and $A(f-s_f)|_{[x_p,x_{p+q+1}]} < m+2q+2$ for every $[x_p, x_{p+q+1}] \subsetneq [a,b]$, $q \geq 0$, characterizes the function from the interior of set of functions in $C[a,b]$ having a strongly unique best approximation from $S_{m,k}$ with k simple active knots. It was proved by Braess [8] that the class of functions $f \in C^{m+1}[a,b]$ with $f^{(m+1)}(t) \neq 0$ for all $t \in (a,b)$ satisfy this property (cf. Johnson [26]).

A necessary alternation condition for unique best approximations from $S_{m,k}$, similar to condition (ii) in Theorem 1.5., was given in [39]. It follows from this result that for every function $f \in C[a,b]$ having a unique best approximation $s_f \in S^1_{m,k}$, we have $A(f - s_f)|_{[a,b]} \geq m + 2k + 2$.

It is well known that local and global best approximations from $S_{m,k}$ are not the same, in general, and that $S_{m,k}$ is not a sun. The same is true for the set $S^1_{m,k}$ (see e.g. Braess [9]). On the other hand, it was shown in [39] that strongly unique local and strongly unique global best approximations from $S^1_{m,k}$ coincide, and that $S^1_{m,k}$ is a sun for strong unicity (i.e. if $s_f \in S^1_{m,k}$ is a strongly unique best approximation of f, then s_f is a strongly unique best approximation of all functions on the ray from s_f through f). Moreover, a general characterization of suns for strong unicity was given in [39].

Next, we consider the set-valued metric projection $P_{S_{m,k}}$ which associates to each $f \in C[a,b]$, the set of best approximations of f from $S_{m,k}$. The following result was proved in Nürnberger [38].

Theorem 1.6. *The set-valued metric projection $P_{S_{m,k}}$ is upper semicontinuous if and only if $k \leq m$, but never lower semicontinuous.*

In the case of fixed knots, it was proved by Nürnberger & Sommer [48] that there exists a continuous selection for $P_{S_m(x_1,\ldots,x_k)}$ if and only if $k \leq m + 1$. (For more general results see the surveys of Deutsch

[18] and Nürnberger & Sommer [50] and the book of Wu Li [29].) However, the problem of the existence of continuous selections for $P_{S_{m,k}}$ is unsolved at present.

Best approximation by generalized splines with free knots was investigated by Nürnberger, Schumaker, Sommer & Strauss [46]. Aside from existence and conditions for best approximations, it was proved that for functions from the convexity cone, the best approximations from continuously composed Haar spaces must be differentiable.

Finally, we discuss the computation of best approximations from $S_{m,k}$. Numerical examples show that the real power of splines is given when the k free knots are located efficiently. Newton type methods, if they converge, only yield local best approximations, in general.

In the following, we describe an algorithm for computing good global best approximations from $S_{m,k}$ which converges if we start with an arbitrary set of knots (Meinardus, Nürnberger, Sommer & Strauss [31]).

We denote by $PP_{m,k}$ the set of *piecewise polynomials of degree m with k free knots.* (For piecewise polynomials, no continuity is required at the knots.) For a subinterval I of $[a,b]$, we set $d(f,\Pi_m,I) = \inf_{p\in\Pi_m} \|(f-p)|_I\|$.

Let a function $f \in C[a,b]$ be given. We compute a spline approximation in two steps:

In the first step, we compute a *leveled set of knots* $\{\bar{x}_1,\ldots,\bar{x}_k\}$ for f, i.e. $a = \bar{x}_0 \leq \bar{x}_1 \leq \cdots \leq \bar{x}_k \leq \bar{x}_{k+1} = b$ such that $d(f,\Pi_m,[\bar{x}_{i-1},\bar{x}_i]) = d(f,\Pi_m,[\bar{x}_i,\bar{x}_{i+1}])$, $i = 1,\ldots,k$. It follows from Lawson [28] that this set is an optimal set of knots for f. Moreover, this set of knots reflects the "critical" parts of f. Because, if the knot-interval $[\bar{x}_i,\bar{x}_{i+1}]$ is relatively small (respectively large), then f is difficult (respectively easy) to approximate on this interval.

In the second step, we compute a (differentiable) best spline approximation of f from $S_m(\bar{x}_1,\ldots,\bar{x}_k)$ by applying the Remez type algorithm already discussed in this section.

By using the inequality $d(f,PP_{m,k}) \leq d(f,S_{m,k}) \leq d(f,S_m(\bar{x}_1,\ldots,\bar{x}_k))$, we get an estimate of the unknown difference of the last two values by the known difference of the first and the third value. (For example, for the function $f(t) = t^{1/2}$ on $[0,2]$, we obtain $d(f,PP_{m,k}) = 0.7\cdot 10^{-4}$ and $d(f,S_m(\bar{x}_1,\ldots,\bar{x}_k)) = 2.6\cdot 10^{-4}$, while $d(f,S_m(x_1,\ldots,x_k)) = 2.3\cdot 10^{-2}$ for equidistant knots. Further examples are given in [37, pp. 157, 158].)

Now, we briefly describe the basic principle of an algorithm for computing a leveled set of knots for f and $PP_{m,k}$ due to Nürnberger, Sommer & Strauss [51].

Algorithm for $PP_{m,k}$

For every $n \geq 1$, we compute a set of knots $\{x_{1,n},\ldots,x_{k,n}\}$ and set $d_{i,n} = d(f,\Pi_m,[x_{i,n},x_{i+1,n}])$, $i = 0,\ldots,k$, where $x_{0,n} = a$ and $x_{k+1,n} = b$, as follows. In the first step, we choose a set of knots $\{x_{1,1},\ldots,x_{k,1}\}$ and compute the values $d_{0,1},\ldots,d_{k,1}$ by applying the classical Remez algorithm (see e.g. Meinardus [30]). We set $a_1 = 10^{\alpha_1} = \min\{d_{i,1} : i = 0,\ldots,k\}$ and $b_1 = 10^{\beta_1} = \max\{d_{i,1} : i = 0,\ldots,k\}$. For $n \geq 1$, we proceed by induction as follows. We set $\delta_{n+1} = (\alpha_n + \beta_n)/2$ and $d_{n+1} = 10^{\delta_{n+1}}$. Then we determine a set of knots $\{x_{1,n+1},\ldots,x_{k,n+1}\}$ such that $d_{i,n+1} = d_{n+1}$, $i = 0,\ldots,k-1$, and compute $c_{n+1} = d_{k,n+1}$. Then we set $a_{n+1} = 10^{\alpha_{n+1}} = \max\{a_n,\min\{d_{n+1},c_{n+1}\}\}$ and $b_{n+1} = 10^{\beta_{n+1}} = \min\{b_n,\max\{d_{n+1},c_{n+1}\}\}$. We successively compute the knots $x_{1,n+1},\ldots,x_{k,n+1}$ as follows. For computing the first knot $x_{1,n+1}$, we choose points y_1 and y_2 such that $d(f,\Pi_m,[a,y_1]) = 10^{\gamma_1} \leq d_{n+1}$ and $d(f,\Pi_m,[a,y_2]) = 10^{\gamma_2} > d_{n+1}$. Then we apply the regula falsi method for the exponents $\gamma_1,\delta_{n+1},\gamma_2$ by setting $y_3 = y_1 + (y_2 - y_1)(\delta_{n+1} - \gamma_1)/(\gamma_2 - \gamma_1)$. In this way, we obtain a sequence (y_p) converging to the knot $x_{1,n+1}$. Then we continue this method.

It was shown in [51] that for $d(f, PP_{m,k}) = 10^\delta$, we have $|\delta - \delta_n| \le \frac{1}{2}|\alpha_{n-1} - \beta_{n-1}| \le \cdots \le \frac{1}{2^{n-1}}|\alpha_1 - \beta_1|$ for all n, and therefore, $\lim_{n\to\infty} d_n = d(f, PP_{m,k})$.

Instead of computing the minimal deviation of f from Π_m on the knot-intervals considered, it is possible to replace this value by simpler expressions resulting e.g. from the first step of the Remez algorithm or from interpolation. For more details, the reader is referred to de Boor [4], Meinardus, Nürnberger, Sommer & Strauss [31], Nürnberger [37], Sommer [69], Grothmann & Mhaskar [21] and the references therein.

2 Interpolation by Bivariate Spline Spaces

In this section, we investigate bivariate spline spaces of the following type. First, the space of *bivariate polynomials of total degree m* is denoted by $\widetilde{\Pi}_m = span\{x^i y^j : i \ge 0,\ j \ge 0,\ i + j \le m\}$. Let a rectangle $T = [a, b] \times [c, d]$ and points $a = x_0 < x_1 < \cdots < x_p = b$, $c = y_0 < y_1 < \cdots < y_q = d$, be given. By defining $R_{i,j} = (x_{i-1}, x_i) \times (y_{j-1}, y_j)$, $i = 1, \ldots, p$, $j = 1, \ldots, q$, we obtain a partition of T into subrectangles $R_{i,j}$, denoted by Δ^0. If the diagonal from (x_{i-1}, y_{j-1}) to (x_i, y_j) is added to each subrectangle $R_{i,j}$, then we denote the resulting partition by Δ^1. Moreover, if we add to each subrectangle $R_{i,j}$ both diagonals, then the resulting partition is denoted by Δ^2.

The spline spaces are defined as follows. Let integers r and m such that $0 \le r \le m$ be given. The space of all functions $f \in C^r(T)$ such that the restriction to each subset of the partition is in $\widetilde{\Pi}_m$ is denoted by $S^r_m(\Delta^\mu)$, $\mu = 0, 1, 2$. These spaces are called spaces of *bivariate splines of degree m and smoothness r* with respect to the partition Δ^μ, $\mu = 0, 1, 2$.

The dimension and basis of bivariate spline spaces in the case of arbitrary triangulations is only known for splines of special degree and smoothness. In general, the dimension depends on the geometry of the underlying partition (see the survey of Schumaker [66] and the book of Chui [11] and the references therein).

On the other hand, the dimension and basis is known in the case of cross-cut partitions, in particular, for uniform partitions. Therefore, in the following we assume that a uniform partition is given (i.e. the knots x_i and y_j which define the partition are uniformly distributed). Moreover, because of space limitations, we only consider the space $S^r_m = S^r_m(\Delta^1)$, although Theorem 2.1. and the construction of interpolation points described below analogously hold for the other partitions, for special non-uniform partitions and for simply connected domains.

The dimension of spline spaces for uniform partitions was determined by Chui & Wang [14] and Schumaker [65]. We begin with a result on the basis of such spaces which is due to Chui & Wang [14] and Dahmen & Micchelli [15], [16].

Theorem 2.1. *A basis of S^r_m is given by the following functions:*

(i) *The polynomials $x^i y^j$, $i + j \le m$.*

(ii) *The truncated power functions $(a_\mu + b_\mu x + c_\mu y)_+^{r+1}\, x^i y^j$, $i + j \le m - r - 1$, where $a_\mu + b_\mu x + c_\mu y = 0$ are the interior grid lines of the partition.*

(iii) *Certain cone splines, easily defined by univariate B-splines (see below).*

The splines in (iii) of Theorem 2.1. can be defined as follows. For each interior grid point of the partition, for simplicity say $z = (0,0)$, we denote by C the union of the cones in $\{(x,y) \in \mathbb{R}^2 : x \geq 0\}$ formed by the three lines of the partition passing through z. We choose a line l which intersects these three lines in $z^0, z^1, z^2 \in C$. For $n = m, m-1, \ldots$, we consider the univariate spline spaces on l, defined by $S_n^r(l) = \{s \in C^r(l) : s|_{[z^i, z^{i+1}]} \in \Pi_n,\ i = 0,1;\ s = 0 \text{ on } l\backslash[z^0, z^2]\}$ until $\dim S_n^r(l) = 0$. Now, we extend each univariate B-spline B in $S_n^r(l)$ to $\mathbb{R}^2$ as follows. Let $B = 0$ on $\mathbb{R}^2\backslash C$. For each ray in C passing through z, we define B to be the univariate truncated power function t_+^{n+1} multiplied with an appropriate constant. The constant is easily determined by requiring that the graph of B on the ray intersects the graph of the univariate B-spline on l. This yields a bivariate spline B in $S_n^r \subset S_m^r$ which is zero outside the cone C.

Now, all bivariate splines in S_m^r obtained in this way are exactly the cone splines in (iii) of Theorem 2.1.. Obviously, the total number of such splines is $(p-1)(q-1) \sum\limits_{n \leq m} \dim S_n^r(l)$.

We now investigate interpolation by bivariate spline spaces. In contrast to the univariate case, it is a non-trivial problem to construct any points at which interpolation by S_m^r is possible. Therefore, we formulate the following problem: Determine points $x^1, \ldots, x^N$ in T, where $N = \dim S_m^r$, such that for all real numbers $\alpha_1, \ldots, \alpha_N$, the *Lagrange interpolation problem* $s(x^i) = \alpha_i$, $i = 1, \ldots, N$, has a unique solution s in S_m^r. Such points $x^1, \ldots, x^N$ are called *Lagrange interpolation points* for S_m^r, and we say that the set $\{x^1, \ldots, x^N\}$ *admits unique Lagrange interpolation* from S_m^r.

Lagrange interpolation points were constructed by Chui, He & Wang [13], Chui & He [12] and Jeeawock-Zedek & Sablonnière [24] for splines of degree at most two. Moreover, Hermite type interpolation points were given by Sha [67], [68], Guo, Sha & Wu [22] and Jeeawock-Zedek [23] for splines of degree at most four. In their approaches, the above authors used Bernstein-Bézier techniques.

In the following, we describe a completely different general method for constructing Lagrange interpolation points for S_m^r which was developed by Nürnberger & Riessinger [41], [42]. (Riessinger [58] investigated our approach in further details for quadratic splines.) We obtain some typical configurations of interpolation points. (More general configurations are given in [41], [42].) Obviously, the construction of interpolation points is equivalent to the construction of total points in the following sense. We call points $x^1, \ldots, x^N$ in T *total points* for S_m^r if $s(x^i) = 0$, $i = 1, \ldots, N$, implies $s = 0$ on T for all $s \in S_m^r$.

In order to describe our method, it suffices to consider the following partition.

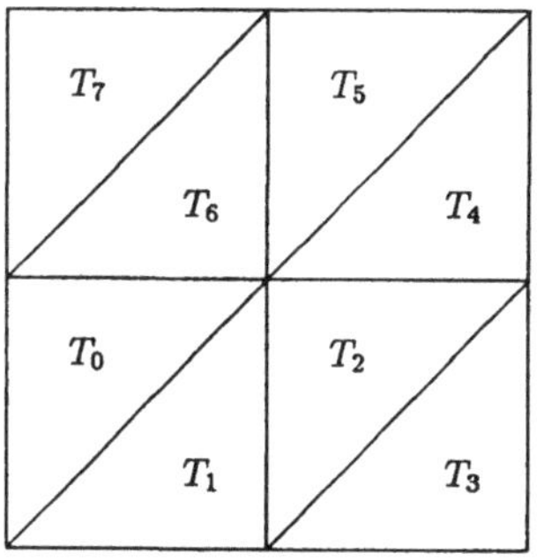

Our basic principle is that for each i, we choose a set of total points for the space $S_{m,i}^r = \{s \in S_m^r : s = 0 \text{ on } T_\mu,\ \mu = 0, \ldots, i-1\}|_{T_i}$ by applying simple rules based on univariate considerations. This is done as follows. Up to some modifications, we choose a line $a + bx + cy = 0$ through $\bigcup\limits_{\mu \leq i} T_\mu$ and total points for $S_{m,i}^r$ restricted to this line. If a spline $s \in S_{m,i}^r$ vanishes on these points, then $s = 0$ on this line and therefore, $s(x,y) = (a + bx + cy)\tilde{s}(x,y)$ with $\tilde{s} \in S^r_{m-1,i}$. By successive degree reduction, we can show that $s = 0$ on $\bigcup_{\mu \leq i} T_\mu$. This is done in [42] by using special bivariate basis functions of $S^r_{m,i}$ (compare also Theorem 2.1.). In this paper, we gave a general interpolation result for spaces of bivariate splines supported in certain cones.

Moreover, we showed that $S^r_{m,i}$ restricted to a given line is actually not only a subspace but the whole univariate spline space which is associated with the line crossing the given partition. The resulting univariate spline spaces are spanned by B-splines, and total points for these spaces are characterized by the well-known interlacing condition of Schoenberg & Whitney (cf. Theorem 1.1.).

Construction of interpolation points

In the following steps, we speak of a (finite) line l through given triangles T_i, $i \in I$, if $l \subset \bigcup_{i \in I} T_i$ and $l \cap T_i$, $i \in I$, are nondegenerate intervals.

In order to construct interpolation points for S^r_m, we have to describe four basic steps. In the first step, we use a set of total points for $\tilde{\Pi}_m$ which is well known.

Step 0. Choose $m+1$ horizontal lines $a_1, \ldots, a_{m+1}$ through T_0. For $j = 1, \ldots, m+1$, choose $\dim S^r_{m-j+1}|_{a_j} = \dim \tilde{\Pi}_{m-j+1}|_{a_j} = m - j + 2$ points on a_j.
Step 1. Choose $m - r$ horizontal lines $b_1, \ldots, b_{m-r}$ through T_0, T_1. For $j = 1, \ldots, m - r$, choose $\dim\{s \in S^r_{m-j+1} : s = 0 \text{ on } T_0\}|_{b_j} = m - j + 1 - r$ points on $b_j \backslash T_0$.
Step 2. Let γ be the largest integer not greater than $m - \frac{3}{2}r$. Choose horizontal lines $c_1, \ldots, c_\gamma$ through T_5. For $j = 1, \ldots, \gamma$, choose $\dim\{s \in S^r_{m-j+1} : s = 0 \text{ on } T_i,\ i = 0, \ldots, 4\}|_{l \cap T_5} = (2(m - j + 2) - 3(r + 1))_+ - ((m - j + 2) - 2(r + 1))_+$ points on $c_j \cap \operatorname{int} T_5$, where l is a line through T_0, T_4, T_5, T_6.
Step 3. Set $\delta := ((m+1) - 2(r+1))_+$. Choose δ diagonal lines $d_1, \ldots, d_\delta$ through T_6. For $j = 1, \ldots, \delta$, choose $\dim\{s \in S^r_{m-j+1} : s = 0 \text{ on } T_i,\ i = 0, \ldots, 5\}|_l = ((m - j + 2) - 2(r + 1))_+$ points on $d_j \cap \operatorname{int} T_6$.

For an arbitrary partition Δ^1, the construction of interpolation points by applying the above steps is as follows. First, we apply Step 0 to the first triangle of the lower row of the partition. Then passing from left to right, we apply Step 1 to all triangles in the lower row and to the last triangle in the next row (by choosing vertical lines in this triangle). Then passing from right to left, we alternatingly apply Step 2 and Step 3 to all triangles in this row and Step 1 to the first triangle in this row. Then we continue this method by considering the next row, again passing from right to left and so on until all triangles of T are considered.

Remark. For each $j = 1, 2, \ldots$, the number of points chosen in Step 1 is equal to the number of univariate truncated power functions of degree $m - j + 1$ and smoothness r. The number of points chosen in Step 2 is equal to the number of univariate B-splines of degree $m - j + 1$ and smoothness r on l with support having nonempty intersection with the (open) interval $l \cap T_5$. The number of points chosen in Step 3 is equal to the number of univariate B-splines of degree $m - j + 1$ and smoothness r on $l \cap T_6$.

In these steps, we consider for $j = 1, 2, \ldots$, univariate spline spaces of degree $m - j + 1$ and smoothness r until the dimension of the spaces is zero.

If $2m > 3r + 1$, then the grid points of the partition may be chosen among the interpolation points. The number of grid points is greater than the dimension of S^r_m for sufficently large p and q if $2m \leq 3r + 1$.

Theorem 2.2. *The set of points constructed by the above method admits unique Lagrange interpolation by S^r_m. In particular, the number of points is equal to the dimension of S^r_m.*

It follows from Theorem 2.2. and the steps described above that the dimension of the bivariate spline space is equal to the sum of the dimensions of certain univariate spline spaces.

From the numerical point of view, it is desirable that computing interpolating splines one has to solve several small systems of linear equations instead of one large system. The above approach has this property, since we pass from one triangle to the next. Meanwhile, this approach was extended by Adam [1] to interpolation by bivariate splines on cross-cut partitions.

Finally, we discuss the approximation order of the spline interpolation methods. We say that an interpolation method for S^r_m has the approximation order $\alpha \leq m+1$ if for each $f \in C^{m+1}(T)$, the interpolating spline satisfies $\|f-s\| \leq K\|D^{m+1}f\|h^\alpha$, where h is the maximal length of the subrectangles of the partition and $K > 0$ is a constant independent of f and h.

In the above mentioned papers on bivariate splines of low degree, the authors proved good approximation order of their interpolation methods. (For general results on the approximation order from bivariate spline spaces S^r_m, the reader is referred to de Boor & Höllig [5], [6], [7], Dahmen & Micchelli [17], Jia [25] and the book of Chui [11].) Recently, the following theorem was proved by Nürnberger [40] for Hermite type configurations of interpolation points which result as a limit case of the Lagrange type configurations, constructed above.

Theorem 2.3. *For special Hermite type configurations of interpolation points, the approximation order of interpolation by S^1_m is m if $m = 4$, and $m+1$ if $m \geq 5$.*

In the case $m = 4$, the approximation order 4 was proved by Guo, Sha & Wu [22] for a completely different configuration of interpolation points, where the difference is that in [40], the interpolating spline can be computed locally by passing from one triangle to the next.

Finally, we give a numerical example. We interpolate the so-called Franke's testfunction $f(x,y) = \frac{3}{4}e^{-\frac{(9x-2)^2+(9y-2)^2}{4}} + \frac{3}{4}e^{-\frac{(9x+1)^2}{49}+\frac{9y+1}{10}} + \frac{1}{2}e^{-\frac{(9x-7)^2+(9y-3)^2}{4}} - \frac{1}{5}e^{-(9x-4)^2+(9y-7)^2}$ by S^1_4 on $T = [0,1] \times [0,1]$. For a suitable configuration of Lagrange interpolation points (which is "close" to a Hermite type configuration), the interpolation error is e.g. $8.4 \cdot 10^{-4}$ using 1173 interpolation points (see Nürnberger & Riessinger [42]).

References

[1] Adam M., Bivariate spline interpolation on convex sets with cross-cut partitions, preprint.

[2] Arndt H., On the uniqueness of best spline approximations with free knots, J. Approx. Theory 11 (1974), 118–125.

[3] Berens H., Nürnberger G., Nonuniqueness and selections in spline approximation, Constr. Approx. 6 (1990), 181–193.

[4] de Boor C., *A Practical Guide to Splines*, Springer, New York, 1978.

[5] de Boor C., Höllig K., Bivariate box splines and smooth pp functions on a three-direction mesh, J. Comp. Appl. Math. 9 (1983), 13–28.

[6] de Boor C., Höllig K., Approximation power from bivariate C^1-cubics: a counterexample, Proc. Amer. Math. Soc. 87 (1983), 649–655.

[7] de Boor C., Höllig K., Approximation power of smooth bivariate pp functions, Math. Z. 197 (1988), 343–363.

[8] Braess D., Chebyshev approximation by spline functions with free knots, Numer. Math. 17 (1974), 357–366.

[9] Braess D., *Nonlinear Approximation Theory*, Springer, Berlin, 1986.

[10] Brosowski B., Deutsch F., Nürnberger G., Parametric approximation, J. Approx. Theory 29 (1980), 261–277.

[11] Chui C.K., *Multivariate Splines*, CBMS-SIAM, Philadelphia, 1988.

[12] Chui C.K., He T.X., On location of sample points in C^1 quadratic bivariate spline interpolation, in: Collatz L., Meinardus G., Nürnberger G. (eds.), *Numerical Methods of Approximation Theory*, ISNM 81, Birkhäuser, Basel, 1987, 30–43.

[13] Chui C.K., He T.X., Wang R.H., Interpolation by bivariate linear splines, in: Szabados J., Tandori K. (eds.), *Alfred Haar Memorial Conference*, North Holland, Amsterdam, 1986, 247–255.

[14] Chui C.K., Wang R.H., Multivariate spline spaces, J. Math. Anal. Appl. 94 (1983), 197–221.

[15] Dahmen W., Micchelli C.A., Recent progress in multivariate splines, in: Chui C.K., Schumaker L.L., Ward J.D. (eds.), *Approximation Theory IV*, Academic Press, New York, 1983, 27–121.

[16] Dahmen W., Micchelli C.A., On the linear independence of multivariate B-Splines II: Complete configurations, Math. Comp. 41 (1983), 143–163.

[17] Dahmen W., Micchelli C.A., On the optimal approximation rates for criss-cross finite element spaces, J. Comp. Appl. Math. 10 (1984), 255–273.

[18] Deutsch F., An exposition of recent results on continuous metric selections, in: Collatz L., Meinardus G., Nürnberger G. (eds.), *Numerical Methods of Approximation Theory*, ISNM 81, Birkhäuser, Basel, 1987, 30–43.

[19] DeVore R., Degree of nonlinear approximation, in: Chui C.K., Schumaker L.L., Ward J.D. (eds.), *Approximation Theory VI*, Academic Press, New York, 1989, 175–201.

[20] Freyburger K., Approximation mit verallgemeinerten Splines, Dissertation, Universität Mannheim, 1991.

[21] Grothmann R., Mhaskar H.N., Detection of singularities using segment approximation, Math. Comp. 59 (1992), 533–540.

[22] Guo Z., Sha Z., Wu Z., On interpolation by splines (Chinese), Chin. Ann. Math. 8 (1987), 368–376.

[23] Jeeawock-Zedek F., Interpolation scheme by C^1 cubic splines on a non uniform type-2 triangulation of a rectangular domain, preprint.

[24] Jeeawock-Zedek F., Sablonnière P., Lagrange and Hermite interpolation by bivariate quadratic splines on nonuniform type 2 triangulations, talk at the conference "Curves and Surfaces", Chamonix, 1993.

[25] Jia R.Q., Approximation order from certain spaces of smooth bivariate splines on a three-direction mesh, Trans. Amer. Math. Soc. 295 (1986), 199–212.

[26] Johnson R.S., On monosplines of least deviation, Trans. Amer. Math. Soc. 96 (1960), 458–477.

[27] Kawasaki H., A second-order property of spline functions with one free knot, J. Approx. Theory, to appear.

[28] Lawson C.L., Characteristic properties of the segmented rational minimax approximation problem, Numer. Math. 6 (1964), 293–301.

[29] Li W., *Continuous Selections for Metric Projections and Interpolating Spaces*, Peter Lang Verlag, Frankfurt, 1991.

[30] Meinardus G., *Approximation of Functions: Theory and Numerical Methods*, Springer, Berlin, 1967.

[31] Meinardus G., Nürnberger G., Sommer M., Strauss H., Algorithms for piecewise polynomials and splines with free knots, Math. Comp. 53 (1989), 235–247.

[32] Mulansky B., Chebyshev approximation by spline functions with free knots, IMA J. Numer. Anal. 12 (1992), 95–105.

[33] Mulansky B., Necessary conditions for local best Chebyshev approximations by splines with free knots, in: Braess D., Schumaker L.L. (eds.) *Numerical Methods of Approximation Theory*, ISNM 105, Birkäuser, Basel, 1992, 195–206.

[34] Nürnberger G., A local version of Haar's theorem in approximation theory, Numer. Funct. Anal. Optim. 5 (1982), 21–46.

[35] Nürnberger G., Best approximation by spline functions: theory and numerical methods, in: Meinardus G., Nürnberger G. (eds.), *Delay Equations, Approximation and Application*, ISNM 74, Birkhäuser, Basel, 1985, 180–212.

[36] Nürnberger G., Strongly unique spline approximation with free knots, Constr. Approx. 3 (1987), 31–42.

[37] Nürnberger G., *Approximation by Spline Functions*, Springer, Berlin, 1989.

[38] Nürnberger G., The metric projection for free knot splines, J. Approx. Theory 71 (1992), 145–153.

[39] Nürnberger G., Strong unicity in nonlinear approximation and free knot splines, Constr. Approx., to appear.

[40] Nürnberger G., Approximation order of bivariate spline interpolation, in preparation.

[41] Nürnberger G., Riessinger T., Lagrange and Hermite interpolation by bivariate splines, Numer. Funct. Anal. Optim. 13 (1992), 75–96.

[42] Nürnberger G., Riessinger T., Bivariate spline interpolation at grid points, preprint.

[43] Nürnberger G., Schumaker L.L., Sommer M., Strauss H., Interpolation by generalized splines, Numer. Math. 42 (1983), 195–212.

[44] Nürnberger G., Schumaker L.L., Sommer M., Strauss H., Generalized Tchebycheffian splines, SIAM J. Math. Anal. 15 (1984), 790–804.

[45] Nürnberger G., Schumaker L.L., Sommer M., Strauss H., Approximation by generalized splines, J. Math. Anal. Appl. 108 (1985), 466–494.

[46] Nürnberger G., Schumaker L.L., Sommer M., Strauss H., Uniform approximation by generalized splines with free knots, J. Approx. Theory 59 (1989), 150–169.

[47] Nürnberger G., Singer I., Uniqueness and strong uniqueness of best approximations by spline subspaces and other spaces, J. Math. Anal. Appl. 90 (1982), 171–184.

[48] Nürnberger G., Sommer M., Characterization of continuous selections of the metric projection for spline functions, J. Approx. Theory 22 (1978), 320–330.

[49] Nürnberger G., Sommer M., A Remez type algorithm for spline functions, Numer. Math. 41 (1983), 117–146.

[50] Nürnberger G., Sommer M., Continuous selections in Chebyshev approximation, in: Brosowski B., Deutsch F. (eds.), *Parametric Optimization and Approximation*, ISNM 72, Birkhäuser, Basel, 1985, 248–263.

[51] Nürnberger G., Sommer M., Strauss H., An algorithm for segment approximation, Numer. Math. 48 (1986), 463–477.

[52] Pinkus A., *On L^1-Approximation*, Cambridge Tracts in Mathematics 93, Cambridge University Press, Cambridge, 1989.

[53] Pinkus A., Strauss H., Best approximation with coefficient constraints, IMA J. Numer. Anal. 8 (1988), 1–22.

[54] Popov V.A., Nonlinear multivariate approximation, in: Chui C.K., Schumaker L.L., Ward J.D. (eds.), *Approximation Theory VI*, Academic Press, New York, 1989, 519–536.

[55] Powell M.J.D., *Approximation Theory and Methods*, Cambridge University Press, 1981.

[56] Rice J.R., Characterization of Chebyshev approximation by splines, SIAM J. Math. Anal. 4 (1967), 557–567.

[57] Rice J.R., *The Approximation of Functions II*, Addison & Wesley, Reading, 1969.

[58] Riessinger T., Interpolation by bivariate quadratic splines on a four-directional mesh, Computing 49 (1992), 129–137.

[59] Schaback R., On alternation numbers in nonlinear Chebyshev approximation, J. Approx. Theory 23 (1978), 379–391.

[60] Schoenberg I.J., Whitney A., On Polya frequency functions III. The positivity of translation determinants with application to the interpolation problem by spline curves, Trans. Amer. Math. Soc. 74 (1953), 246–259.

[61] Schumaker L.L., Uniform approximation by Tchebycheffian spline functions, J. Math. Mech. 18,1 (1968), 369–378.

[62] Schumaker L.L., Uniform approximation by Chebyshev spline functions II: free knots, SIAM J. Numer. Anal. 5 (1968), 647–656.

[63] Schumaker L.L., On the smoothness of best spline approximations, J. Approx. Theory 2 (1969), 410–418.

[64] Schumaker L.L., *Spline Functions: Basic Theory*, Wiley-Interscience, New York, 1981.

[65] Schumaker L.L., Bounds on the dimension of spaces of multivariate piecewise polynomials, Rocky Mountain J. Math. 14 (1984), 251–264.

[66] Schumaker L.L., On spaces of piecewise polynomials in two variables, in: Singh S.P., Burry J.H.W., Watson B. (eds.), *Approximation Theory and Spline Functions*, Reidel, Dodrecht, 1984, 151–197.

[67] Sha Z., On interpolation by $S_2^1(\Delta_{m,n}^2)$, Approx. Theory and Its Applic. 1 (1985), 71–82.

[68] Sha Z., On interpolation by $S_3^1(\Delta_{m,n}^1)$, Approx. Theory and Its Applic. 1 (1985), 1–18.

[69] Sommer M., Segment approximation using linear functionals, in: Braess D., Schumaker L.L. (eds.) *Numerical Methods of Approximation Theory*, ISNM 105, Birkäuser, Basel, 1992, 331–346.

[70] Sommer M., Strauss H., Weak Descartes systems in generalized spline spaces, Constr. Approx. 4 (1988), 133–145.

[71] Sommer M., Strauss H., Order of strong uniqueness in best L_∞-approximation by spline spaces, Acta Math. Hung. 61 (1993), 259–280.

[72] Strauss H., Eindeutigkeit bei der gleichmäßigen Approximation mit Tschebyscheffschen Splinefunktionen, J. Approx. Theory 15 (1975), 78–82.

[73] Watson G.A., *Approximation Theory and Numerical Methods*, Wiley-Interscience, Chichester, 1980.

2nd Int. Coll. on Numerical Analysis, pp. 155-164
D. Bainov and V. Covachev (Eds)

Parallelizable PCG Methods for the Cray Y-MP and the TMC CM-2

Thomas C. Oppe, William H. Holter, I. M. Navon

Supercomputer Computations Research Institute
Florida State University, Tallahassee, FL, USA

Abstract

This paper presents the results of applying a number of vectorizable and parallelizable preconditioned conjugate gradient methods to the numerical solution of the diffusion equation governing the flow of groundwater in a confined two-dimensional rectangular aquifer. Four different sample problems are formulated having coefficients (transmissivities) ranging from continuous and isotropic to sharply discontinuous and anisotropic. The preconditioners applied to these problems include the Jacobi, line Jacobi, incomplete LU (ILU) and modified incomplete LU (MILU) decompositions, symmetric Gauss-Seidel, least squares polynomial, and a new alternating direction preconditioner. The suitability of each of these preconditioning methods for the Y-MP and Connection Machine 2 (CM-2) parallel computers is investigated. It is shown that the ILU and MILU methods are effective on a single-processor Y-MP when vectorized using a wavefront strategy. For the CM-2, the least squares polynomial and the alternating direction preconditioners are the most effective of the methods used in the study.

1 Introduction

This paper provides a comparison of the performance characteristics of several preconditioned conjugate gradient (PCG) methods as implemented on the Cray Y-MP and the Thinking Machines Corporation CM-2. These methods are used for solving four example problems arising from the modeling of groundwater flow via discretization of the 2-D diffusion equation. Varying properties of the groundwater flow model, such as isotropic vs. anisotropic and continuous vs. discontinuous transmissivities, give rise to large sparse systems of linear equations whose matrices can be very ill-conditioned.

For this study, traditional preconditioners that vectorize and parallelize well but have poor convergence rates, including Jacobi, line Jacobi, least squares polynomial, and symmetric Gauss-Seidel with red-black ordering are implemented on both computers. Preconditioning methods that parallelize poorly but have good convergence rates, such as ILU(k) and MILU(k) with natural ordering, are implemented on the Y-MP employing various optimization strategies. Finally, a new preconditioner resembling the Alternating Direction Implicit (ADI) method is developed in order to achieve increased parallelism while maintaining the good convergence rate enjoyed by the MILU preconditioner.

The paper is organized as follows. Section 2 contains a description of each of the four test problems in terms of their respective diffusion equation parameters and boundary conditions; Sections 3 and 4 contain a description of the PCG algorithms used in the study; Section 5 presents an evaluation of the numerical results; and, lastly, Section 6 presents some concluding remarks.

2 The Test Problems

The flow of groundwater in a confined aquifer may be described by the 2-D diffusion equation:

$$s\frac{\partial u}{\partial t} = \frac{\partial}{\partial x}\left(a\frac{\partial u}{\partial x}\right) + \frac{\partial}{\partial y}\left(b\frac{\partial u}{\partial y}\right) + f, \tag{1}$$

where s is the storage coefficient of the porous media, u is the pressure head (height of the water table) to be determined, a and b are (non-negative) transmissivities in the x and y directions, respectively, and f is a source/sink term that takes into account recharging/discharging wells.

Each of the test problems is defined on the unit square whose sides are assigned appropriate boundary conditions. A steady-state solution is sought for each problem, in which case Eq. (1) simplifies to:

$$-\frac{\partial}{\partial x}\left(a\frac{\partial u}{\partial x}\right) - \frac{\partial}{\partial y}\left(b\frac{\partial u}{\partial y}\right) = f. \tag{2}$$

The boundary conditions considered are no-flow (homogeneous Neumann) and constant head (Dirichlet). Discretizing on a uniform node-centered grid with distance between adjacent nodes $\Delta x = \Delta y = h$, Eq. (2) becomes:

$$-\alpha_{i-1,j}u_{i-1,j} - \alpha_{i,j}u_{i+1,j} - \beta_{i,j-1}u_{i,j-1} - \beta_{i,j}u_{i,j+1} + \sigma_{i,j}u_{i,j} = h^2 f_{i,j} = g_{i,j}, \tag{3}$$

where $i = 1, 2, \ldots, n_x$ and $j = 1, 2, \ldots, n_y$, with n_x and n_y denoting the number of grid points in the x and y directions, respectively. The α and β coefficients, located halfway between nodes, are given respectively by the harmonic means of the a and b coefficients at adjacent nodes in order to ensure continuity of fluid flux.

For an arbitrary rectangular uniform grid, the resulting system of equations for the pressure heads, u, may be written as

$$Au = b,$$

where the matrix A is symmetric and positive definite (SPD) and the vector b is a function of g and the boundary conditions. The matrix A has a pentadiagonal structure and is weakly diagonally dominant.

The four problems considered in this paper are described below.

2.1 EXPNA

This problem has continuous and isotropic transmissivities, a and b, and fully Dirichlet boundary conditions. The transmissivities are

$$a(x, y) = b(x, y) = 100(x + y).$$

The forcing function f and the boundary conditions are chosen so that the analytic solution to Eq. (2) is

$$u(x, y) = \cos(4\pi x)\cos(4\pi y).$$

2.2 EXPNC

This problem has continuous and anisotropic transmissivities, a and b, and fully Dirichlet boundary conditions. The transmissivities are

$$\begin{aligned} a(x,y) &= 100x, \\ b(x,y) &= 100(1-y). \end{aligned}$$

The forcing function f and the boundary conditions are chosen so that the analytic solution to Eq. (2) is the same as that given for EXPNA.

2.3 EXP10G

This problem has discontinuous and isotropic transmissivities, a and b, and fully Dirichlet boundary conditions. The values of the transmissivities range from 0 (impervious) to 100,000 (highly permeable) and are piecewise constant in certain subdomains as given in Figure 1. In this figure, filled circles with negative g values indicate discharging wells and hollow circles with positive g values indicate recharging wells.

2.4 EXP6G

This problem has discontinuous and anisotropic transmissivities, a and b, and mixed Neumann and Dirichlet boundary conditions. The values of the transmissivities range from 0 to 100,000 and are piecewise constant in certain subdomains as given in Figure 2. In this figure $\delta = 0$ indicates a no-flow boundary condition while $u = 0$ indicates a Dirichlet boundary condition.

3 The PCG Method

Let Q be a symmetric and positive definite matrix, called the *preconditioning matrix*, such that $Q \approx A$ and the solution to $Q\delta = r$ given r is "easy" to compute. The Preconditioned Conjugate Gradient (PCG) algorithm used in this study is given by:

For $n = 0, 1, 2, \ldots$ until convergence do

$$\begin{aligned} r^{(n)} &= \begin{cases} b - Au^{(0)} & n = 0 \\ r^{(n-1)} - \alpha_{n-1} z^{(n-1)} & n \geq 1 \end{cases} \\ & \quad \text{If } \|r^{(n)}\|_2 / \|b\|_2 < \zeta, \text{ exit} \\ \delta^{(n)} &= Q^{-1} r^{(n)} \\ \gamma_n &= \langle \delta^{(n)}, r^{(n)} \rangle \\ \beta_n &= \gamma_n / \gamma_{n-1} \qquad (\beta_0 = 0) \\ p^{(n)} &= \delta^{(n)} + \beta_n p^{(n-1)} \qquad (p^{(-1)} = 0) \\ z^{(n)} &= Ap^{(n)} \\ \alpha_n &= \gamma_n / \langle p^{(n)}, z^{(n)} \rangle \\ u^{(n+1)} &= u^{(n)} + \alpha_n p^{(n)} \end{aligned}$$

where $r^{(n)}$ are the residuals, $u^{(n)}$ are the iterates, and ζ is the convergence criterion. Note that the speed with which the algorithm converges is strongly dependent on the choice of Q. In general, it is desirable that the condition number of $Q^{-1}A$ be smaller than that of A.

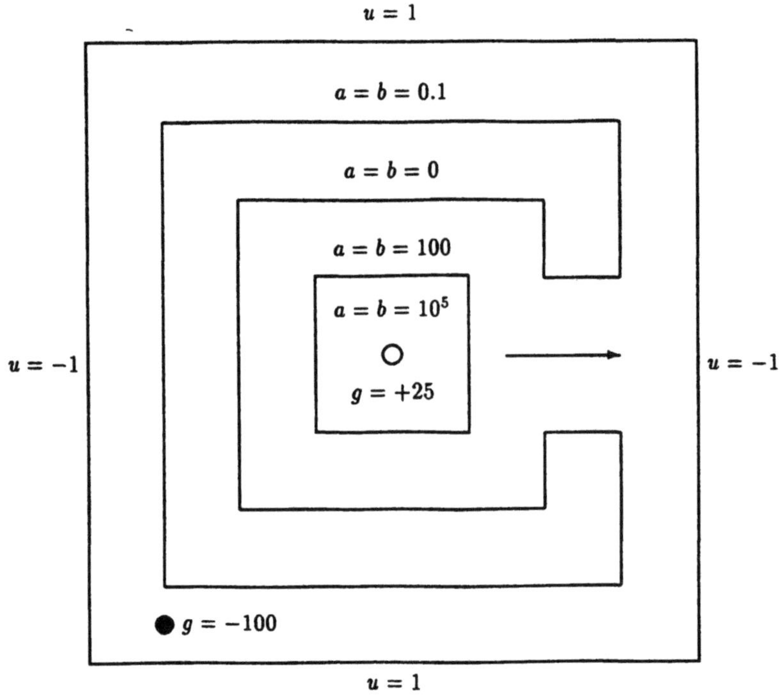

Figure 1: EXP10G

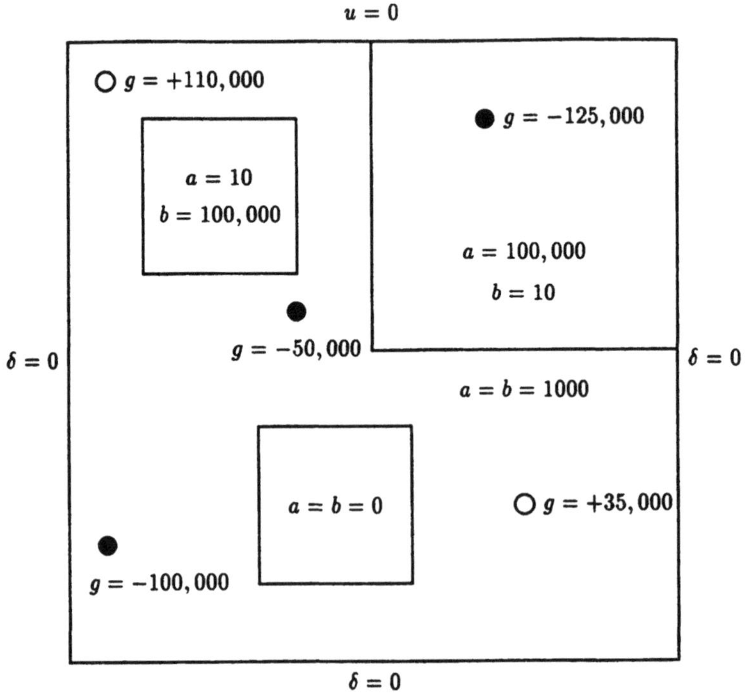

Figure 2: EXP6G

Many components of the PCG algorithm can be computed with optimized assembly language routines on both the Y-MP and the CM-2. For the Y-MP, the Basic Linear Algebra Subprograms (BLAS)[1] routines are used to compute inner product, 2-norm, and vector update operations. For the CM-2, the Connection Machine Scientific Software Library (CMSSL) [2] routines are used to compute inner product, 2-norm, and matrix-vector product operations.

The matrix A is symmetrically scaled to have a unit diagonal with the transformation

$$\left(D^{-\frac{1}{2}}AD^{-\frac{1}{2}}\right)\left(D^{\frac{1}{2}}u\right) = \left(D^{-\frac{1}{2}}b\right),$$

where D is the main diagonal of the unscaled matrix. On the Y-MP, the east and north coefficients are stored by diagonals in one-dimensional vectors. This allows the matrix-vector product $z = Ap$ to be computed with long vector operations. On the CM-2, the matrix coefficients are stored as two-dimensional arrays.

4 Preconditioning Techniques

Let the diagonally scaled matrix A be written as

$$A = I + E + N + W + S,$$

where the E, N, W, and S matrices contain the east, north, west, and south coefficients, respectively. For the symmetric problems considered in this study, $W = E^T$ and $S = N^T$. The preconditioning methods used in the study are described below in terms of these matrices.

4.1 Jacobi

For this method, Q is the main diagonal of A. For the scaled matrix, the main diagonal is I, and thus the preconditioning step is

$$\delta = Q^{-1}r = r.$$

This operation is trivially vectorizable and parallelizable on both the Y-MP and CM-2.

4.2 Line Jacobi

For this method, $Q = T$, where T is a tridiagonal matrix formed from A. We consider two cases:

line Jacobi, x direction: In this case, $T = I + W + E$, so that T is formed from the east and west coefficients. Note that T is composed of n_y independent tridiagonal systems of length n_x.

line Jacobi, y direction: In this case, $T = I + N + S$, so that T is formed from the north and south coefficients. Note that T is composed of n_x independent tridiagonal systems of length n_y.

In either case, the preconditioning step,

$$\delta = T^{-1}r,$$

involves solving independent tridiagonal systems of equal size. This operation vectorizes easily on the Y-MP by operating on corresponding elements of all the systems at once. For the CM-2, specially optimized routines in the CMSSL (version 3.0) library for factoring and solving independent tridiagonal systems are used.

4.3 Symmetric Gauss-Seidel (Red-Black Ordering)

If A is permuted to have a red-black structure, then A can be written as

$$A = \begin{pmatrix} I & F_R \\ F_B & I \end{pmatrix}.$$

In this case, the symmetric Gauss-Seidel (SGS) preconditioner is given by

$$Q = \begin{pmatrix} I & 0 \\ F_B & I \end{pmatrix} \begin{pmatrix} I & F_R \\ 0 & I \end{pmatrix}.$$

The preconditioning step $\delta = Q^{-1} r$ is given by the two-step process

$$\begin{aligned} \delta_B &= r_B - F_B r_R \\ \delta_R &= r_R - F_R \delta_B \end{aligned}$$

These operations vectorize and parallelize well. On the Y-MP, constant-stride vector operations of length $\frac{1}{2}n$, where n is the system size, are used on the unpermuted matrix and vectors. On the CM-2, the optimized CMSSL matrix-vector product routine is used in conjunction with logical mask arrays to update the black points first, followed by an update of the red points.

4.4 Incomplete *LU* Decomposition

The ILU method is very popular on scalar computers and has been studied extensively [3, 4, 5, 6, 7, 8]. For this method, an incomplete LU decomposition of A is used for Q. Thus,

$$Q = (M + L)\, M^{-1}\, (M + U),$$

where M, L, and U are diagonal, strictly lower triangular, and strictly upper triangular matrices, respectively.

The methods used in the study are the unmodified method, ILU(k), and the fully modified method, MILU(k), where $k \geq 0$ denotes the level of fill-in of nonzeros. Typically, Q more closely approximates A with increasing values of k but also has greater storage costs.

This study investigates the use of the ILU and MILU methods for various levels of fill-in under the natural ordering of the unknowns and considers various techniques for vectorizing the forward and backward solution steps on the Y-MP. The ILU and MILU methods are not competitive on the CM-2 due to the recursiveness of these operations. For the Y-MP, two approaches are tried for optimizing the forward and backward solution steps:

Line-Oriented: With this approach, described in [9], each line of nodes is updated one at a time. There is an inherent recursion between the nodes of a single line. Cray Assembly Language (CAL) routines are used to implement these recursions.

Computational Wavefronts: In this approach, described in [10, 11], the nodes are grouped into computational wavefronts in which all nodes belonging to the same wavefront can be updated simultaneously. For the ILU(0) pattern, the computational wavefronts are the grid diagonals running from northwest to southeast. Unfortunately, the vector lengths are much shorter than the problem size, and increasing the level of fill-in results in even smaller vector lengths.

4.5 Symmetric Alternating Direction Implicit (SADI)

Let

$$A = D + E + N + W + S,$$

where D, E, N, W, and S are the diagonal, east, north, west, and south coefficients, respectively. Let

$$H = D_H + E + W \qquad V = D_V + N + S$$

for diagonal matrices D_H and D_V chosen so that $D = D_H + D_V$ and H and V are positive semi-definite. Thus, $A = H + V$, where H is tridiagonal, and V can be permuted to be tridiagonal. Let Ω and Ω' be diagonal matrices with positive elements, and define the four-step basic iterative method:

$$\begin{aligned}
(H+\Omega)\,u^{(n+\frac{1}{4})} &= (-V+\Omega)\,u^{(n)}+b\\
(V+\Omega')\,u^{(n+\frac{1}{2})} &= (-H+\Omega')\,u^{(n+\frac{1}{4})}+b\\
(V+\Omega')\,u^{(n+\frac{3}{4})} &= (-H+\Omega')\,u^{(n+\frac{1}{2})}+b\\
(H+\Omega)\,u^{(n+1)} &= (-V+\Omega)\,u^{(n+\frac{3}{4})}+b.
\end{aligned}$$

This method can also be written as

$$\begin{aligned}
u^{(n+\frac{1}{2})} &= G_1 u^{(n)} + k_1\\
u^{(n+1)} &= G_2 u^{(n+\frac{1}{2})} + k_2,
\end{aligned}$$

where $G_i = I - Q_i^{-1}A$, $k_i = Q_i^{-1}b$ for $i = 1, 2$, and

$$\begin{aligned}
Q_1 &= (H+\Omega)\,(\Omega+\Omega')^{-1}\,(V+\Omega')\\
Q_2 &= (V+\Omega')\,(\Omega+\Omega')^{-1}\,(H+\Omega).
\end{aligned}$$

The splitting matrix corresponding to this composite iterative method is given by

$$Q = Q_1\,(Q_1 + Q_2 - A)^{-1}\,Q_2.$$

Note that if A is symmetric, $Q_2 = Q_1^T$ so $Q^T = Q$. If Q is also SPD, conjugate gradient acceleration can be applied to the basic iterative method.

The preconditioning operation $\delta = Q^{-1}r$ involves solving tridiagonal systems corresponding to $H + \Omega$ and $V + \Omega'$. On the Y-MP, these tridiagonal systems are solved in vector mode by treating corresponding elements of the tridiagonal subsystems as vectors. On the CM-2, special CMSSL software for solving multiple independent tridiagonal systems is used.

A crucial issue for this method is the selection of the diagonal matrices Ω and Ω'. Suppose first that $\Omega = \omega I$ and $\Omega' = \omega' I$. Suppose also that the eigenvalues of H lie in the interval $[a, b]$, and the eigenvalues of V lie in the interval $[c, d]$. Then a good choice of the parameters ω and ω', in the sense that a bound on the spectral radius of the ADI iteration matrix is minimized [12], is given by:

$$\begin{aligned}
\omega &= \left[-ab + cd + \sqrt{(a+c)(a+d)(b+c)(b+d)}\right] \Big/ (a+b+c+d)\\
\omega' &= \left[ab - cd + \sqrt{(a+c)(a+d)(b+c)(b+d)}\right] \Big/ (a+b+c+d)
\end{aligned}$$

Using eigenvalue bounds for H and V, these values for ω and ω' are used for problem EXPNA. For the remaining problems, these values resulted in Q being indefinite. Good values for ω and ω' were found by experimentation for EXPNC and EXP10G, but they could not be found for EXP6G.

4.6 Least Squares Polynomial

For this method, the preconditioning matrix Q is such that

$$Q^{-1} = p_n(A)$$

where p_n is an n-th degree polynomial. The polynomials $\{p_n\}$ are to be chosen so that $Q^{-1}A \approx I$ in a least squares sense. Specifically, p_n is chosen so that

$$\int_m^M [1 - xp_n(x)]^2 \, w(x)\, dx$$

is minimized, where w is a positive weighting function and the interval $[m, M]$ contains the spectrum of A. For this study, the weighting function

$$w(x) = x^c(M - x)^d$$

is used where $c > -1$ and $d > -1$ are chosen to emphasize portions of the spectrum of A. Thus $c = d = 0$ results in a uniform weighting of the spectrum, while $c = d = -\frac{1}{2}$ results in a greater weighting of the extremal eigenvalues than of the interior eigenvalues. For the four test problems considered in this paper, the choice $c = d = -\frac{1}{2}$ results in a more effective preconditioner than the choice $c = d = 0$. The preconditioning step

$$\delta = Q^{-1}r = p_n(A)r$$

is evaluated using Horner's rule and repeated matrix-vector multiplication operations.

5 Performance Analysis and Observations

The PCG methods discussed in Section 4 were applied to the four test problems described in Section 2. For all runs, the starting solution is $u^{(0)} = 0$ and the stopping criterion is

$$\frac{\|r^{(n)}\|_2}{\|b\|_2} < 10^{-6}.$$

The mesh sizes used for the problems are $n_x = n_y = 64$, $n_x = n_y = 128$, and $n_x = n_y = 256$ for the CM-2 and slightly smaller for the Y-MP to avoid memory bank conflicts.

The performance results are tabulated in [13], and the reader is referred to this report for a more detailed description of the experiment.

5.1 Y-MP Results

Based on the results obtained on the Cray Y-MP, the following observations are noted:

1. The most effective methods for the problems with continuous coefficients (EXPNA and EXPNC) are SADI and the wave version of MILU(1). The most effective methods for the problems with discontinuous coefficients (EXP10G and EXP6G) are, respectively, SADI and the wave version of ILU(k) with high values of k. MILU(k) cannot be used for either problem since Q is not positive definite.
2. For the ILU(k) and MILU(k) methods, the wave approach is faster than the line approach except for the smallest problem size with high values of k.
3. The red-black method SGS-RB performs better than Jacobi by halving the number of iterations and using long vector operations.

4. The least squares method is more effective than the Jacobi method. Higher degree polynomials are more effective than lower degree polynomials up to some optimal degree.
5. For problem EXPNA, the SADI method performs very well with an iteration count close to that of MILU(3). Using good values for the iteration parameters ω and ω', the number of iterations grows according to $h^{-\frac{1}{2}}$ as does MILU(k). For EXP6G, the SADI method is not as effective since good values could not be found.

5.2 CM-2 Results

Similarly, the following observations are noted based on the results obtained on the CM-2:

1. The SADI method is the most effective method for the problems with continuous coefficients for the largest problem size. For the problems with discontinuous coefficients or for small problem sizes, the least squares polynomial method is the most effective. As noted previously, the ILU and MILU methods were not competitive on the CM-2 due to the recursiveness of the forward and backward solution steps.
2. The least squares method is significantly faster than the Jacobi method, indicating that the inner product calculations are expensive compared to computing matrix-vector products. Higher degree polynomials are more effective than lower degree polynomials up to some optimal degree.

6 Concluding Remarks

This study has investigated the performance characteristics of a number of alternative preconditioners incorporated into the PCG method and applied to the solution of the two-dimensional diffusion equation on the Y-MP and the CM-2. The implementations of the resulting algorithms were optimized for both machines and subsequently utilized in solving four sample problems of varying degrees of complexity—coefficients ranging from continuous and isotropic to sharply discontinuous and anisotropic—for each of three increasingly larger problem sizes.

The selection of the "best" preconditioner is strongly dependent upon the complexity of the problem at hand, the size of the problem, and the computer being utilized. Although the study indicates that the selection cannot be performed *a priori*, the observations given below should prove useful in the selection process for similar problems.

For continuous problems on the Y-MP, SADI and the wave versions of MILU(k) perform better than their competitors, primarily because the number of iterations required for their convergence grows like $h^{-\frac{1}{2}}$ rather than h^{-1}. For discontinuous problems, SADI is preferred if "good" values of ω and ω' can be found; otherwise, the wave versions of ILU(k) are preferable since Q for MILU(k) is not positive definite.

Neither the ILU(k) nor MILU(k) preconditioners parallelize well due to their inherent recursiveness; hence, they are not effective on the CM-2. LSP(n) (for $n \approx 10$) is most effective on that machine for "small" problems, but is eventually surpassed by SADI (if applicable).

Acknowledgements

This research was supported in part by the Florida State University Supercomputer Computations Research Institute which is partially funded by the U.S. Department of Energy through Contract No. DE-FC05-85ER250000. Time on the Cray Y-MP was provided in part by the Florida State University Computation Center.

References

[1] C. L. Lawson, R. J. Hanson, D. R. Kincaid, and F. T. Krogh. "Basic Linear Algebra Subprograms for Fortran Usage." *ACM Transactions on Mathematical Software*, Vol. 5, No. 3, September 1979, pp. 308–323.

[2] *CMSSL Release Notes, Version 3.0 Beta*, Thinking Machines Corporation, Cambridge, Massachusetts, January, 1992.

[3] T. Dupont. "A Factorization Procedure for the Solution of Elliptic Difference Equations." *SIAM Journal of Numerical Analysis*, Vol. 5, No. 4, December 1968, pp. 753–782.

[4] T. Dupont, R. P. Kendall, and H. H. Rachford. "An Approximate Factorization Procedure for Solving Self-Adjoint Elliptic Difference Equations." *SIAM Journal of Numerical Analysis*, Vol. 5, No. 3, September 1968, pp. 559–573.

[5] I. Gustafsson. "A Class of First Order Factorization Methods." *BIT*, Vol. 18, 1978, pp. 142–156.

[6] I. Gustafsson. *Stability and Rate of Convergence of Modified Incomplete Cholesky Factorization Methods.* Doctoral dissertation, Chalmers University of Technology and the University of Göteborg, April 1979.

[7] D. S. Kershaw. "The Incomplete Cholesky–Conjugate Gradient Method for the Iterative Solution of Systems of Linear Equations." *Journal of Computational Physics*, Vol. 26, No. 1, January 1978, pp. 43–65.

[8] J. A. Meijerink and H. A. van der Vorst. "An Iterative Solution Method for Linear Systems of Which the Coefficient Matrix is a Symmetric M-Matrix." *Mathematics of Computation*, Vol. 31, No. 137, January 1977, pp. 148–162.

[9] A. Greenbaum and G. H. Rodrigue. "The Incomplete Cholesky Conjugate Gradient Method for the STAR (5-point operator)." Research Report UCID-17574, Lawrence Livermore Laboratory, Livermore, CA., 1977.

[10] C. Ashcraft. "A Moving Computation Front Approach for Vectorizing ICCG Calculations." General Motors Research Publication, GMR-5174, 1985.

[11] C. Ashcraft and R. Grimes. "On Vectorizing Incomplete Factorizations and SSOR Preconditioners." *SIAM Journal on Scientific and Statistical Computing*, Vol. 9, No. 1, January 1988, pp. 122–151.

[12] D. M. Young. *Iterative Solution of Large Linear Systems.* New York: Academic Press, 1971.

[13] W. H. Holter, I. M. Navon, and T. C. Oppe. "Parallelizable Preconditioned Conjugate Gradient Methods for the Cray Y-MP and the TMC CM-2." Research Report FSU-SCRI-92-128, Supercomputer Computations Research Institute, Florida State University, September 1992.

2nd Int. Coll. on Numerical Analysis, pp. 165-176
D. Bainov and V. Covachev (Eds)

Stability of Finite Difference Formulas for Linear Parabolic Equations

OLE ØSTERBY
Computer Science Department, Aarhus University
DK 8000 Aarhus C, Denmark

Abstract - The notion of stability is often confusing because several concepts hide under one name. In this paper we study a number of examples from the literature involving linear parabolic equations with constant coefficients with emphasis on cases where disagreements occur and attempt to explain the reasons for the differences.

Keywords - Parabolic Equations, Difference Formulas, Stability.
MSC - 65M12

We shall be concerned with initial-boundary value problems for the equation

$$u_t = u_{xx} - \phi u_x - \kappa u \qquad 0 \le x \le a,\ 0 \le t$$

This type of problems arises frequently in physics and chemistry, where the equation $u_t = u_{xx}$ models diffusion, ϕu_x is a transport term, and κu models a (chemical) reaction. Especially when ϕ or κ are large compared to the diffusion coefficient which is set to 1 here, numerical methods may have difficulties.

The initial condition is $u(x,0) = f(x)$, $0 \le x \le a$,
and the boundary conditions are

$$\begin{aligned} a_0 u(0,t) - b_0 u_x(0,t) &= g_0(t) \qquad & 0 \le t \\ a_1 u(a,t) + b_1 u_x(a,t) &= g_1(t) \qquad & 0 \le t \end{aligned}$$

where a_0, a_1, b_0, b_1 are non-negative, and $a_0 + b_0 > 0$ and $a_1 + b_1 > 0$. If $b_0 = b_1 = 0$ we have Dirichlet boundary conditions, and if $a_0 = a_1 = 0$ we have Neumann boundary conditions and otherwise we have boundary conditions of Robin type or of the third kind. We note that the sign conventions will guarantee that the solution stays bounded for $t > 0$, a condition which is usually required for simulation of a physical system.

As for the equation itself, if $\kappa = 0$ we have an advection-diffusion problem with limiting cases being pure diffusion ($u_t = u_{xx}$), or pure advection $u_t = \phi u_x$, and if $\phi = 0$ we have a diffusion-reaction system.

We shall study numerical solutions to the parabolic equation using various finite-difference methods. Several of these can be derived in a two-stage process, where we first discretize in space using

$$\begin{aligned} h^2 u_{xx}(x,t) &\approx \delta^2 u_n^m &= u_{n+1}^m - 2u_n^m + u_{n-1}^m \\ 2h u_x(x,t) &\approx 2\mu\delta u_n^m &= u_{n+1}^m - u_{n-1}^m \end{aligned}$$

where h and k are the step sizes in the x- and t-direction respectively, and $x = nh$ and $t = mk$, and u_n^m is an approximation to the value of the true solution u at the point (x,t). With $U = (u_0, u_1, \ldots, u_N)^T$, $Nh = a$ the problem is now transformed into a linear system of ordinary differential equations:

$$\frac{dU}{dt} = \frac{1}{h^2}(AU + s).$$

The matrix A can be written as

$$A = T - \frac{1}{2}\phi h S - \kappa h^2 I.$$

In case of Dirichlet boundary conditions it is customary to leave u_0 and u_N out of the vector U. The matrices A, T and S are now of dimension $N-1$ and

$$T = \left\{ \begin{array}{cccc} -2 & 1 & & \\ 1 & -2 & 1 & \\ & 1 & \cdot & \cdot \\ & & \cdot & \cdot \end{array} \right\}, \qquad S = \left\{ \begin{array}{cccc} 0 & 1 & & \\ -1 & 0 & 1 & \\ & -1 & \cdot & \cdot \\ & & \cdot & \cdot \end{array} \right\}.$$

If the boundary condition at 0 involves the first derivative then $hu_x(0,t)$ can be approximated by $u_1^m - u_0^m$ or by $(-u_2^m + 4u_1^m - 3u_0^m)/2$ or a higher order approximation, and this constitutes the basis for the first row in matrix A. Since symmetric difference approximations are more accurate than asymmetric ones a popular alternative is $2hu_x(0,t) \approx u_1^m - u_{-1}^m$ where the 'fictitious' value u_{-1}^m is eliminated by using the difference approximation for the differential equation also for $x = 0$. In this case u_0 is retained in vector U. Similar considerations apply at the other boundary $x = a$ if the boundary condition here involves u_x.

The properties of the matrix A determine the behaviour of the system of ordinary differential equations and can therefore be used to study the effects of the discretization in the x-direction.

The system of ordinary differential equations can now be solved numerically by the general *θ-method:*

$$U^{m+1} - U^m = \lambda\{\theta A U^{m+1} + (1-\theta)AU^m + \theta s^{m+1} + (1-\theta)s^m\}$$

where $U^m = (u_0^m, u_1^m, \ldots, u_N^m)^T$, $\lambda = k/h^2$, and $\theta \in [0,1]$. The most popular values for θ are 0, 1, and $\frac{1}{2}$ corresponding to (in ODE terms) Euler, Backward Euler, and the trapezoidal rule, or (in PDE-terms) the explicit method, the fully implicit (or Laasonen) method [8] and the Crank-Nicolson method [3], respectively. Solving for U^{m+1} we get

$$(I - \lambda\theta A)U^{m+1} = (I + \lambda(1-\theta)A)U^m + \lambda(\theta s^{m+1} + (1-\theta)s^m)$$

or

$$U^{m+1} = BU^m + q^m$$

where

$$B = (I - \lambda\theta A)^{-1}(I + \lambda(1-\theta)A)$$

and

$$q^m = \lambda(I - \lambda\theta A)^{-1}(\theta s^{m+1} + (1-\theta)s^m).$$

Our main concern in this paper shall be the stability of numerical methods for parabolic equations, i.e. the property that errors do not amplify as the calculations progress. The propagation of errors is governed by the homogeneous equation, and we shall therefore set $q^m = 0$ in what follows.

Stability is a very important concept, first of all because of the Lax equivalence theorem [14] that stability is necessary and sufficient for convergence when a consistent finite-difference approximation is used to solve a properly posed problem. Convergence here is the property that the numerical solution at a fixed point (x,t) converges to the true solution at (x,t) as the step sizes h and k tend to 0, and convergence thus means, that we can get results as accurate as we wish by using small enough steps. But small steps means many calculations and therefore large computation times, not to mention round-off errors. But the term stability is also associated with the property, that errors are not magnified without bound as calculations with a fixed step size progress. This sounds like two different although related properties and we shall see that this is indeed the case.

A parallel situation exists as regards ordinary differential equations where the term *0-stability* is generally accepted for the former kind of stability, and the term *absolute stability* corresponds to the latter. The prefix *0-* is to remind us of the limiting process for the step sizes. In partial differential equations no generally accepted nomenclature has been adopted to distinguish between the two kinds of stability, although the latter meaning of the word has been associated with the terms *fixed step stability*, *practical stability* or *strict stability*, cf. [10, 11].

A result by Widlund [19] states (among other things) that the stability of a method is independent of the low order terms. This of course is a result about '0-stability' as also indicated by the condition

$$\| U^{m+1} \| \leq (1 + 0(k)) \| U^m \| .$$

For fixed k, or for values of k that may change but are bounded below by some positive constant, we may still encounter exponential growth of the numerical solution and of the error. For problems where the solution is known to stay bounded this is clearly unsatisfactory.

Two methods are in general use for studying the stability of numerical methods for parabolic equations: The von Neumann method [12, 16] and the matrix method, and these are the methods we shall study here with emphasis on examples where they seem to disagree.

The von Neumann method assumes a Fourier-type expansion of the numerical solution and studies the propagation of the individual terms from one time-step to the next. We thus set

$$u_n^m = e^{\alpha mk} e^{i\beta nh} = \gamma^m e^{i\beta nh}.$$

Note that

$$\begin{aligned} \delta^2 u_n^m &= \gamma^m e^{i\beta nh}(e^{i\beta h} - 2 + e^{-i\beta h}) \\ &= 2u_n^m(\cos\beta h - 1) \\ &= -4u_n^m \sin^2 \tfrac{\beta h}{2} \end{aligned}$$

and

$$\begin{aligned} 2\mu\delta u_n^m &= \gamma^m e^{i\beta nh}(e^{i\beta h} - e^{-i\beta h}) \\ &= 2iu_n^m \sin\beta h \end{aligned}$$

where i is the imaginary unit. For the θ-method we now get after division by u_n^m

$$\gamma - 1 = (\theta\gamma + 1 - \theta)(-4\lambda \sin^2 \frac{\beta h}{2} - i\phi \frac{k}{h} \sin \beta h - \kappa k).$$

A method is called von Neumann-stable if the magnification of each individual term in each step $|e^{\alpha k}| = |\gamma| \leq 1$ (or $1 + O(k)$ if we are concerned with '0-stability').

As seen above the von Neumann method is very easy to apply and as we shall see later, it is often capable of giving general results and stability conditions. But the von Neumann method is basically designed for pure initial-value problems with periodic initial values. Since most practical problems are (formulated as) initial-boundary value problems, the question is how the (discretization of the) boundary values influences the stability properties. Here the matrix method proves its value for it studies directly the transformation, as effected by the matrix B above, of the numerical solution from one time step to the next including the boundary conditions as they appear in the matrices A and B.

A sufficient condition for stability is that

$$\| U^m \| < C \| U^0 \|$$

for any m, where C is a constant and $\| \cdot \|$ is a suitable matrix norm. It follows that another sufficient condition for stability is that

$$\| B \| < 1.$$

Since we know that $\rho(B) \leq \| B \|$ for any matrix norm, where $\rho(B)$ denotes the spectral radius of B, i.e. the absolute value of the largest eigenvalue of B, and also that for any $\epsilon > 0$ there exists a matrix norm such that $\| B \| < \rho(B) + \epsilon$ [17], a necessary condition for stability is that

$$\rho(B) \leq 1.$$

Because the matrix method takes the (discretization of the) boundary conditions into account it ought to be more reliable than the von Neumann method, but the norm inequalities often give only sufficient conditions and therefore possibly pessimistic results, and the calculation of eigenvalues is time-consuming and the results may give rise to misinterpretations as we shall see.

In the following we shall look at a number of different examples and study the results as produced by the von Neumann method and by the matrix method.

Example 1: The explicit method ($\theta = 0$) on the diffusion-reaction problem ($\phi = 0$) with initial condition $f(x) = 1$ and Dirichlet boundary conditions $u(0,t) = 0$ and $u(a,t) = \exp(-\kappa t)$ with $a \gg 1$. The von Neumann method gives

$$\gamma = 1 - 4\lambda \sin^2 \frac{\beta h}{2} - \kappa k$$

We note that γ is real and less than 1, and the condition $|\gamma| \leq 1$ is equivalent to $\gamma \geq -1$ or

$$\lambda \leq \frac{1}{2} - \frac{1}{4}\kappa k$$

(cf. [1]). We note a stricter condition than the usual $\lambda \leq \frac{1}{2}$ for pure diffusion if we are concerned with 'practical stability'.

The matrix T has eigenvalues $\mu_j = -4\sin^2 \frac{j\pi}{2N}$, $(j = 1, 2, \ldots, N-1)$ [9], and we can therefore give the eigenvalues of $A = T - \kappa h^2 I$ and of $B = I + \lambda A$. The latter are $\omega_j = 1 - 4\lambda \sin^2 \frac{j\pi}{2N} - \kappa k$, $(j = 1, 2, \ldots, N-1)$. The spectral radius of B is $\rho(B) = -\omega_{N-1}(B) = -1 + \kappa k + 4\lambda \cos^2 \frac{\pi}{2N}$. These results are very similar to von Neumann, but they stress the fact that for finite step sizes only a finite set of frequencies are applicable. Thus we can in Fig. 1 display stable behaviour for the pure diffusion case even though $\lambda > \frac{1}{2}$. We note however that the difference between von Neumann and the matrix method is only appreciable for crude step sizes and therefore of limited practical interest. In Fig. 2 we display unstable behaviour for the diffusion-reaction case even though $\lambda < \frac{1}{2}$ to demonstrate the difference between '0-stability' and practical stability, cf. [2].

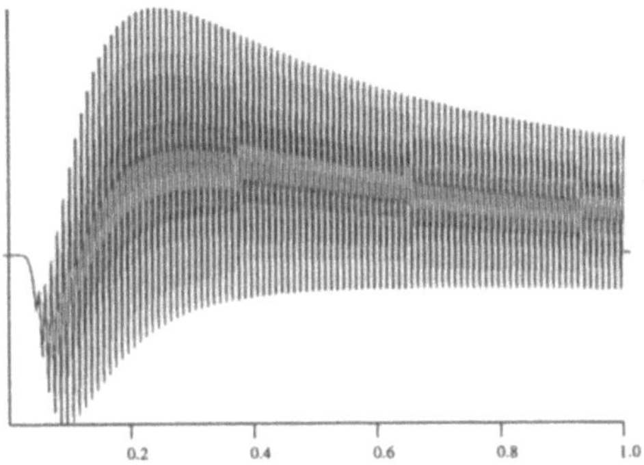

Fig. 1.
Stability of pure diffusion with $\lambda > 0.5$.
Error at $x = 1$ with $h = 0.1$, $\lambda = 0.501$, $\kappa = 0$.

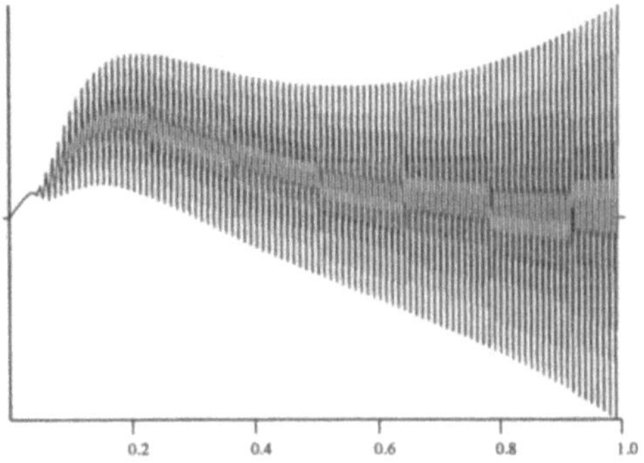

Fig. 2.
Instability of diffusion-reaction with $\lambda < 0.5$.
Error at $x = 1$ with $h = 0.1$, $\lambda = 0.498$, $\kappa = 4.8$.

In this case it is possible for us to give the eigenvalues explicitly. More often we shall have to make do with some matrix norm, such as the 1-norm or the ∞-norm, or with bounds for the spectrum as given by the Gerschgorin circles [17]. In this example the difference is negligible since the eigenvalues spread out rather evenly in the interval $1 - \kappa k - 4\lambda \leq \omega_j \leq 1 - \kappa k$.

Example 2: Consider the explicit method on the advection-diffusion problem ($\kappa = 0$) with initial condition $f(x) = 0$ and boundary conditions $u(0,t) = 1, u_x(1,t) = 0$ as studied by Siemieniuch and Gladwell [15]. A von Neumann analysis gives with $\alpha = \frac{1}{2}\phi h$:

$$\gamma = 1 - 4\lambda \sin^2 \frac{\beta h}{2} - 2i\alpha\lambda \sin \beta h.$$

γ is now complex, but it turns out that the imaginary part has no influence on the 0-stability so the condition for this is still $\lambda \leq \frac{1}{2}$ [11]. For fixed-step stability we must also require $\alpha^2 \lambda \leq \frac{1}{2}$.

The matrix B in this case is

$$B = \left\{ \begin{array}{ccccc} 1-2\lambda & \lambda(1-\alpha) & & & \\ \lambda(1+\alpha) & 1-2\lambda & \lambda(1-\alpha) & & \\ & \lambda(1+\alpha) & \cdot & \cdot & \\ & & \cdot & \cdot & \lambda(1-\alpha) \\ & & & 2\lambda & 1-2\lambda \end{array} \right\}$$

We note that

$$\begin{array}{lllll} \alpha \leq 1 & \Rightarrow & \| B \|_\infty = |1-2\lambda| + 2\lambda & = 1 & if \quad \lambda \leq \frac{1}{2} \\ \alpha > 1 & \Rightarrow & \| B \|_\infty = |1-2\lambda| + 2\lambda\alpha & > 1 & \end{array}$$

so the norm condition for stability is more severe than von Neumann, since it does not allow $\alpha > 1$ at all.

For $\alpha < 1$ the eigenvalues are real and distinct and satisfy

$$1 - \lambda(2 + 2\sqrt{1-\alpha^2}) < \omega_j < 1 - \lambda(2 - 2\sqrt{1-\alpha^2})$$

For $\alpha = 1$, $\omega_j = 1 - 2\lambda$ is a multiple eigenvalue and for $\alpha > 1$ the eigenvalues are complex

$$\omega_j = 1 - 2\lambda - i\lambda\nu_j, \quad -2\sqrt{\alpha^2-1} < \nu_j < 2\sqrt{\alpha^2-1}$$

The conditions for the eigenvalues to be of absolute value less than or equal to one are therefore

$$\begin{aligned} \alpha \leq 1: &\quad (1 + \sqrt{1-\alpha^2})\lambda \leq 1 \\ \alpha > 1: &\quad \alpha^2\lambda \leq 1. \end{aligned}$$

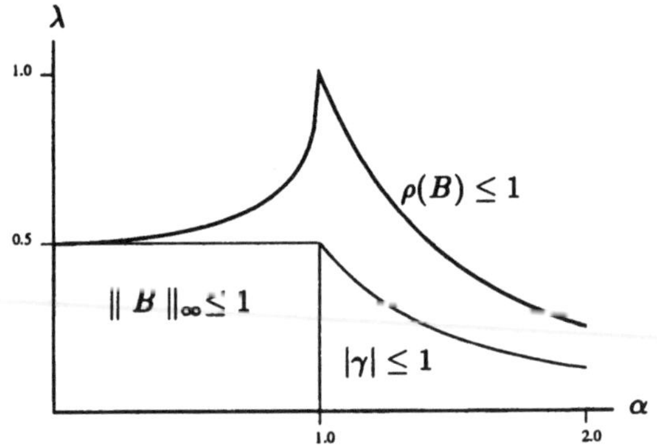

Fig. 3.
Stability regions depending on α and λ
as determined by $\| B \|$, $|\gamma|$ or $\rho(B)$.

As seen in Fig. 3 this is a rather larger stability region, but computations performed by Siemieniuch and Gladwell [15] indicate that especially for large values of ϕ the smaller region seems more correct. In Fig. 4 and Fig. 5 we show the results of calculations with $\phi = 40$, $h = 0.05$ and thus $\alpha = 1$, and with $\lambda = 0.51$ and 0.55 respectively. Already for these comparatively small values of λ (as compared to the spectral radius condition) we notice oscillations which die out as they must but reach considerable amplitudes considering the modest value of ϕ. In Fig. 6 and Fig. 7 we have used $\alpha = 2$ and λ-values corresponding to $\rho(B) = 1$ and $|\gamma| = 1$, respectively. The oscillations die out slowly and are clearly unacceptable in the former case.

Morton [11] blames the matrix method for being misleading in this case (cf. also [5, 6]). A more diplomatic statement might be that Siemieniuch and Gladwell have overinterpreted the matrix method by focusing their attention too much on the eigenvalues. As pointed out by Trefethen [18] the eigenvalues of non-normal matrices are not well suited to characterize the behaviour of the corresponding linear transformations. Also Young [20] has earlier and in quite a different context pointed out that $\| B^m \|$ might grow considerably even though $\rho(B) < 1$. Of course *eventually* $\| B^m \|$ will decay to 0 but this may take a long time or rather a large value of m. In our definition of stability we have allowed for a certain growth by a constant factor C, but if C is several orders of magnitude then the practical value of this stability definition fades away.

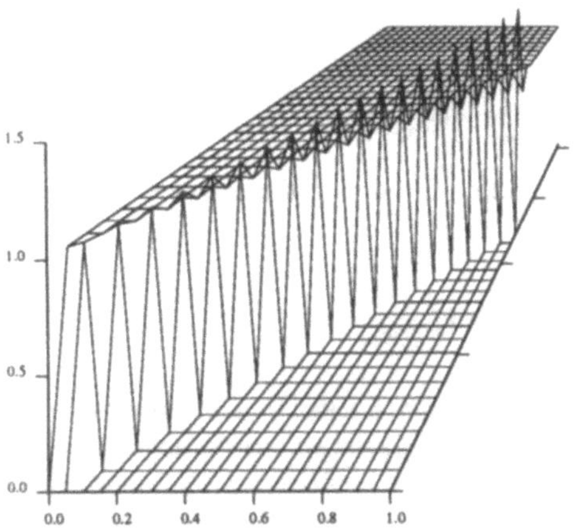

Fig. 4.
Oscillations with amplitude 0.5
with $\alpha = 1$ and $\lambda = 0.51$

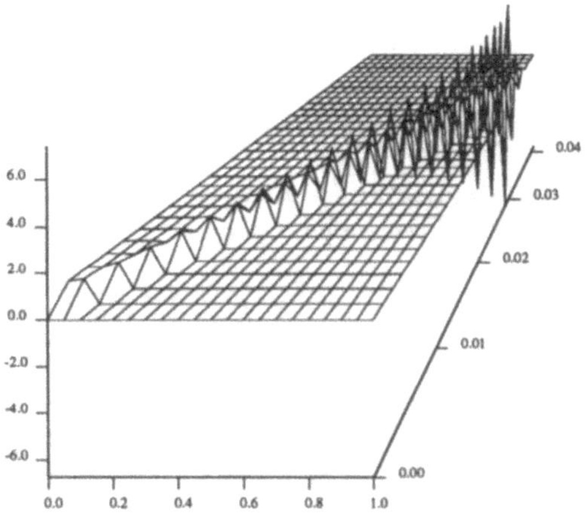

Fig. 5.
Oscillations with amplitude 6
with $\alpha = 1$ and $\lambda = 0.55$

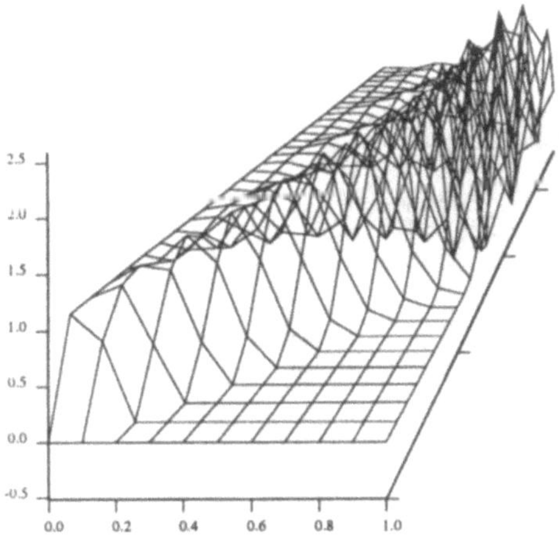

Fig. 6.
Weakly damped oscillations
with $\alpha = 2$ and $\lambda = 0.25$

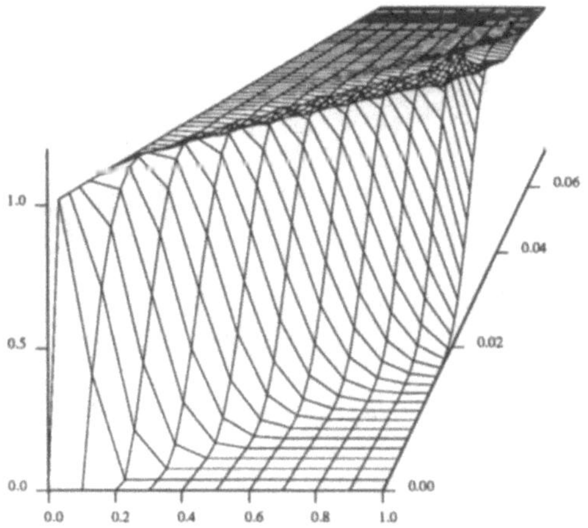

Fig. 7.
Oscillations at the von Neumann stability limit
with $\alpha = 2$ and $\lambda = 0.125$

We know that if $\rho(B) < 1$ then there exists a matrix norm such that $\| B \| < 1$ but this norm may not be suitable for our considerations. In a finite-dimensional vector space all norms are equivalent, but if h is small then the dimension of the vector space is large and the constants involved in the equivalence may be very large and we are back where we started. So what we should learn from this observation is not that the matrix method is misleading, but that eigenvalues of non-normal matrices may be misleading, and that we should be careful with constants in our definitions.

Example 3: For Crank-Nicolson's method on $u_t = u_{xx}$ with Dirichlet boundary data a von Neumann analysis gives

$$\gamma = \frac{1 - 2\lambda \sin^2 \frac{\beta h}{2}}{1 + 2\lambda \sin^2 \frac{\beta h}{2}}$$

and it is seen that γ is real and $-1 \leq \gamma \leq 1$ for arbitrary positive λ.

As for the matrix method

$$B = (I - \frac{1}{2}\lambda A)^{-1}(I + \frac{1}{2}\lambda A).$$

The eigenvectors for B are the same as the eigenvectors for A and T and the eigenvalues for B can therefore be given as

$$\omega_j = \frac{1 + \frac{1}{2}\lambda\mu_j}{1 - \frac{1}{2}\lambda\mu_j}, \qquad \mu_j = -4\sin^2\frac{j\pi}{2N}.$$

Since $\mu_j < 0$ we immediately have $-1 < \omega_j < 1$ proving matrix stability for any positive λ. We note, however, that for very large λ there will be a predominance of eigenvalues very close to -1, giving rise to the well-known 'wiggles' or oscillations associated with the Crank-Nicolson method [4]. A similar analysis for the fully implicit (or Laasonen) method will give $0 < \omega_j < 1$ and a much smoother behaviour of the numerical solution, but this should not be mistaken for accurate values.

Example 4: Same but with a derivative boundary condition at $x = 0$.

$$a_0 u(0,t) - b_0 u_x(0,t) = g_0(t)$$

is discretized

$$a_0 u_0^m - \frac{b_0}{2h}(u_1^m - u_{-1}^m) = g_0^m$$

or

$$u_{-1}^m = -2h\frac{a_0}{b_0}u_0^m + u_1^m - 2h\frac{g_0^m}{b_0}$$

such that

$$u_{-1}^m - 2u_0^m + u_1^m = -2(1 + \rho_0 h)u_0^m + 2u_1^m - 2h\frac{g_0^m}{b_0}$$

where $\rho_0 = a_0/b_0$. The matrix $A = T$ is now of dimension N and is given by

$$A = \begin{Bmatrix} -2(1+\rho_0 h) & 2 & & & & \\ 1 & -2 & 1 & & & \\ & 1 & -2 & . & & \\ & & . & . & . & \\ & & & . & -2 & 1 \\ & & & & 1 & -2 \end{Bmatrix}.$$

Using Gerschgorin's theorem we see that the eigenvalues of A are ≤ 0 when $\rho_0 \geq 0$, and we conclude that the Crank-Nicolson method is stable. A derivative boundary condition at the other end point is treated similarly. Matrix A then has dimension $N + 1$ and the last row contains 2 and $2(1 + \rho_1 h)$, with $\rho_1 = a_1/b_1$. A detailed analysis is given in [7, 9]. Their conclusion is that Crank-Nicolson will become unstable in the presence of derivative boundary conditions, but only if $\rho_0 + \rho_1 + \rho_0\rho_1 < 0$. If $\rho_0 + \rho_1 + \rho_0\rho_1 = 0$ one eigenvalue of A is 0 and consequently one eigenvalue of B is equal to 1 giving rise to a 'persistent error' [13] because the corresponding component of the error vector is not damped at all. In Fig. 8 we show the persistent error at $x = 0.5$ associated with $\rho_0 = \rho_1 = 0$ in Keast and Mitchell's example. In Fig. 9 we have used $\rho_0 = 0$, $\rho_1 = -0.5$ leading to a positive eigenvalue for A and consequently an eigenvalue for B which is larger than 1. We note that the ensuing kind of instability is more difficult to detect in practice than the oscillatory kind associated with negative or complex eigenvalues which we have met in examples 1 and 2.

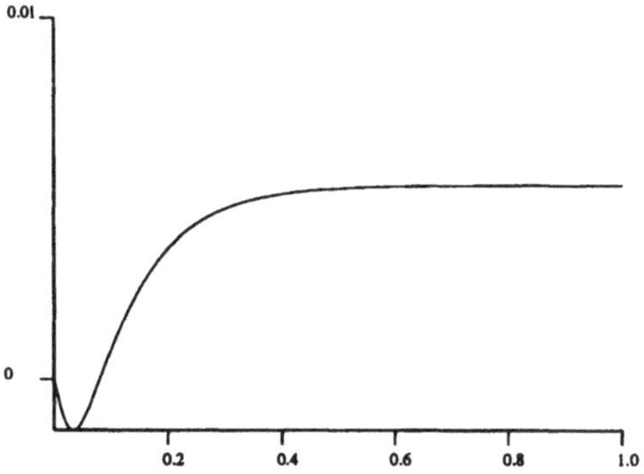

Fig. 8.
Persistent error at $x = 0.5$
with $\lambda = 0.5$, $\rho_0 = 0$, $\rho_1 = 0$

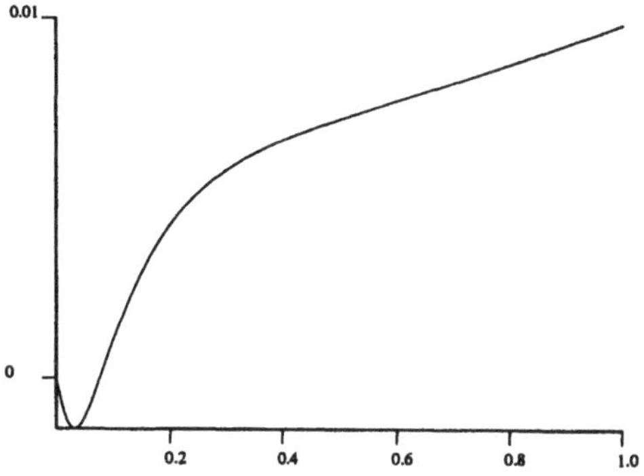

Fig. 9.
Instability of C-N. Error at $x = 0.5$
with $\lambda = 0.5$, $\rho_0 = 0$, $\rho_1 = -0.5$

Example 5: If we use the one-sided difference for discretizing the boundary condition at 0 we have

$$a_0 u_0^m - \frac{b_0}{h}(u_1^m - u_0^m) = g_0^m$$

or

$$u_0^m = \tau_0 u_1^m + \frac{g_0^m}{a_0 + \frac{b_0}{h}}, \qquad \tau_0 = \frac{1}{1+\rho_0 h}.$$

This is used to eliminate u_0^m:

$$u_0^m - 2u_1^m + u_2^m = (\tau_0 - 2)u_1^m + u_2^m + \frac{g_0^m}{a_0 + \frac{b_0}{h}}.$$

Note that $0 < \tau_0 \le 1$ provided $\rho_0 \ge 0$.
The matrix $A = T$ is now of dimension $N-1$ and is given by

$$A = \begin{Bmatrix} \tau_0 - 2 & 1 & & & & \\ 1 & -2 & 1 & & & \\ & 1 & -2 & . & & \\ & & . & . & . & \\ & & & . & -2 & 1 \\ & & & & 1 & -2 \end{Bmatrix}.$$

Again Gerschgorin's theorem guarantees a stable method.
Actually the eigenvalues μ_j of A are given by (cf. [13])

$$\mu_j = -4\sin^2\frac{\psi_j}{2},$$

where ψ_j are the $N-1$ roots in the interval $0 < \psi < \pi$ of

$$\sin N\psi - \tau_0 \sin(N-1)\psi = 0.$$

A detailed inspection (cf. Fig. 10) reveals that

$$0 < \frac{\pi}{2N} < \psi_1 < \frac{\pi}{N} < \cdots < \pi\frac{2N-3}{2N} < \psi_{N-1} < \frac{\pi(N-1)}{N} < \pi.$$

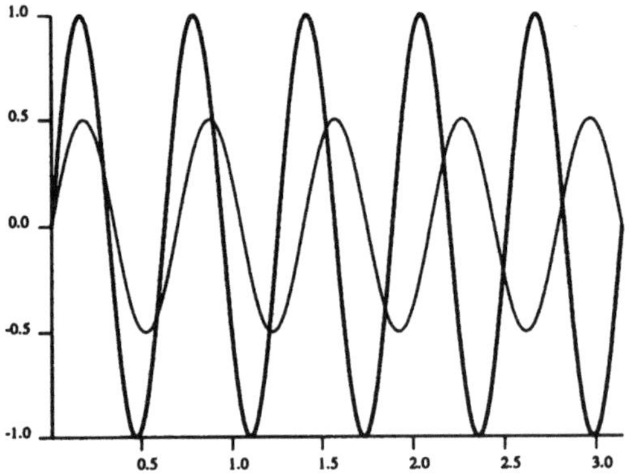

Fig. 10.
Graph of $\sin 10\psi$ and $0.5 \sin 9\psi$

Therefore

$$-4 < -4\cos^2 \frac{\pi}{2N} < \mu_{N-1} < \cdots < \mu_1 < -4\sin^2 \frac{\pi}{4N} < 0.$$

We note that the μ_j are still bounded away from 0 and -4.

Example 0: Time-dependent boundary conditions.
If a_0 and b_0 are functions of t, then $\tau_0 = \tau_0(t)$ may vary with time and

$$B = (I - \frac{1}{2}\lambda A')^{-1}(I + \frac{1}{2}\lambda A)$$

where A' involves $\tau_0' = \tau_0(t+k)$ and A contains $\tau_0(t)$. A' and A need not have the same eigenvectors any more, and the eigenvalues of B are not easily given. We can no longer hope that the damping effect of the first factor can completely compensate for the shortcomings of the second factor if $\lambda > 1$. In fact if we consider

$$(I + \frac{1}{2}\lambda A)v = \omega(I - \frac{1}{2}\lambda A')v$$

and insert $\omega = -1$ and $v = (1, 0, 0, \ldots)^T$ we get from the first component

$$1 + \frac{1}{2}\tau_0\lambda - \lambda = -1 + \frac{1}{2}\tau_0'\lambda - \lambda$$

or

$$(\tau_0' - \tau_0)\lambda = 4,$$

so if τ_0 is an increasing function of t and λ is sufficiently large then B has an eigenvalue $\omega = -1$. If λ is slightly larger then a continuity argument makes it highly plausible that B may possess eigenvalues less than -1, although we have not demonstrated that. Since $0 \le \tau_0 \le 1$ the above condition can only be fulfilled for a limited number of steps, if $\tau(t)$ is monotone which is often the case, but for large values of λ the number might still be large and we are back to the earlier discussion about allowable constants. We note in passing that since this possible instability involves negative eigenvalues it is easy to detect because it will reveal itself by oscillations.

The matrix B can be viewed as a perturbation of

$$B' = (I - \frac{1}{2}\lambda A')^{-1}(I + \frac{1}{2}\lambda A'),$$

the difference being

$$\Delta B = B - B' = \frac{\lambda}{2}(I - \frac{1}{2}\lambda A')^{-1}(A - A').$$

Since B' is a symmetric matrix the eigenvalues are perturbed no more than

$$\| \Delta B \|_2 \leq \frac{\lambda}{2} \| (I - \frac{1}{2}\lambda A')^{-1} \|_2 \| A - A' \|_2 \leq \frac{\lambda}{2}|\tau_0 - \tau_0'|$$

(cf. [17]). The eigenvalues of B' are

$$\omega_j = \frac{1 + \frac{\lambda}{2}\mu_j}{1 - \frac{\lambda}{2}\mu_j},$$

where μ_j are given in the previous example. For large values of λ the largest eigenvalue, ω_1, is clearly smaller than 1, whereas the smallest eigenvalue is

$$\omega_{N-1} = \frac{1 + \frac{\lambda}{2}\mu_{N-1}}{1 - \frac{\lambda}{2}\mu_{N-1}} > \frac{1 - 2\lambda}{1 + 2\lambda} \approx -1 + \frac{1}{\lambda}.$$

Combining we get the condition

$$\lambda^2|\tau_0 - \tau_0'| \leq 2$$

to ensure $\rho(B) \leq 1$, although the assurance is not 100 percent.

Conclusion: In a series of examples we have tried to demonstrate the difference between *0-stability* and *fixed-step stability* and we have tried to explain the different results one might get from using von Neumann's method and the matrix method for investigating the stability of a finite difference scheme.

Acknowledgements: I am indepted to Dieter Britz, Department of Chemistry, Aarhus University and to Leslaw Bieniasz, Molten Salts Laboratory, Krakow, Poland for several enlightening discussions pertaining to the concept of stability.

References

[1] Leslaw K. Bieniasz. The von Neumann stability of finite difference algorithms for the electrochemical kinetic simulation of diffusion coupled with homogeneous reactions. *J. Electroanal. Chem.*, 345:13–25, 1993.

[2] Dieter Britz and Ole Østerby. Some numerical investigations of the stability of electrochemical digital simulation, especially as affected by first-order homogeneous reactions. *J. Electroanal. Chem.*, 1993. in press.

[3] J. Crank and P. Nicolson. A practical method for numerical integration of solution of partial differential equations of heat-conduction type. *Proc. Camb. Philos. Soc.*, 43:50–67, 1947.

[4] Philip M. Gresho and Robert L. Lee. Don't suppress the wiggles – they are telling you something. *Computers and Fluids*, 9:223–253, 1981.

[5] D. F. Griffiths, I. Christie, and A. R. Mitchell. Analysis of error growth for explicit difference schemes in conduction-convection problems. *Internat. J. Numer. Meth. Engnrg.*, 15:1075–1081, 1980.

[6] A. C. Hindmarsh, P. M. Gresho, and D.F. Griffiths. The stability of explicit Euler time-integration for certain finite difference approximations of the multi-dimensional advection-diffusion equation. *Internat. J. Numer. Meth. Fluids*, 4:853–897, 1984.

[7] P. Keast and A. R. Mitchell. On the instability of the Crank-Nicolson formula under derivative boundary conditions. *Comp. J.*, 9:110–114, 1966.

[8] P. Laasonen. Über eine Methode zur Lösung der Wärmeleitungsgleichung. *Acta Math.*, 81:309–317, 1949.

[9] A. R. Mitchell and D. F. Griffiths. *The Finite Difference Method in Partial Differential Equations*. John Wiley, Chichester, 1980.

[10] K. W. Morton. Stability and convergence in fluid flow problems. *Proc. Roy. Soc. Lond. A*, 323:237–253, 1971.

[11] K. W. Morton. Stability of finite difference approximations to a diffusion-convection equation. *Internat. J. Numer. Meth. Engnrg.*, 15:677–683, 1980.

[12] C.G. O'Brien, M.A. Hyman, and S. Kaplan. A study of the numerical solution of partial differential equations. *J. Math. Phys.*, 29:223–251, 1951.

[13] I. B. Parker and J. Crank. Persistent discretization errors in partial differential equations of parabolic type. *Comp. J.*, 7:163–167, 1964.

[14] Robert D. Richtmyer and K. W. Morton. *Difference Methods for Initial-value Problems*, volume 4 of *Tracts in Pure and Applied Mathematics*. Interscience, New York, 2 edition, 1967.

[15] J. L. Siemieniuch and I. Gladwell. Analysis of explicit difference methods for a diffusion-convection equation. *Internat. J. Numer. Meth. Engnrg.*, 12:899–916, 1978.

[16] G. D. Smith. *Numerical Solution of Partial Differential Equations: Finite Difference Methods*. Clarendon Press, Oxford, 2 edition, 1978.

[17] G. W. Stewart. *Introduction to Matrix Computations*. Academic Press, New York, 1973.

[18] L. N. Trefethen. Pseudospectra of matrices. In D. F. Griffiths and G. A. Watson, editors, *Numerical Analysis, Dundee 1991*, pages 234–266. Longman, 1991.

[19] O. B. Widlund. On the stability of parabolic difference schemes. *Math. Comp.*, 19:1–13, 1965.

[20] David M. Young. *Iterative Solution of Large Linear Systems*. Academic Press, New York, 1971.

2nd Int. Coll. on Numerical Analysis, pp. 177-186
D. Bainov and V. Covachev (Eds)

NONLINEAR SPLINES IN CONVEX AND S-CONVEX INTERPOLATION

Jochen W. Schmidt
Institute of Numerical Mathematics
Technical University of Dresden
D - 01062 Dresden, Germany

Abstract. This is a short review of results in convex and S-convex interpolation which includes some recent extensions.
Keywords: Convexity preservation, rational splines, refined grids.

1. INTRODUCTION

On the grid $\Delta_n : x_0 < x_1 < \cdots < x_n$ let the data set

$$D_n = \{(x_i, h_i) : i = 0(1)n\} \tag{1}$$

be given, and assume that D_n is in (strict) convex position, i.e.

$$\tau_1 \leq \tau_2 \leq \cdots \leq \tau_n \ (\tau_1 < \tau_2 < \cdots < \tau_n), \tag{2}$$

where $\tau_i = (h_i - h_{i-1})/\xi_i$ are the slopes and $\xi_i = x_i - x_{i-1}$ the mesh sizes. In convex interpolation we are interested to find a function $s \in C^k$ with $k = 1$, $k = 2$, or even with $k = 3$ which interpolates D_n,

$$s(x_i) = h_i, \ i = 0(1)n, \tag{3}$$

and which, in addition, is convex on the interval $I = [x_0, x_n]$.

Polynomials are unsuited for convex interpolation. Splines are much more promising though there are spline classes which cannot guarantee always success. In the early paper [11] Passow and Roulier give a strictly convex data set D_n, $n \geq 4$ such that all cubic C^1-spline interpolants are not convex on I. In [5] the function $f(x) = |x|$ and the nodes $x_0 = -1$, $x_1 = -0.5$, $x_2 = 0$, $x_3 = 0.5$, $x_4 = 1$ are used in order to demonstrate that, more general, there may be no polynomial C^1-spline interpolants which are convex on $I = [-1, 1]$. Indeed, all interpolating polynomial splines which are convex have to be identical with f, thus they are not from C^1. Further, since the set of data D_n for which the convex interpolation fails is an open set, in a neighbourhood of the above example there are data sets, even in strict convex position, for which convex interpolation with any polynomial C^1 splines is without success. Recently, Mulansky and Neamtu [9] have extended this negative result to the convex interpolation with respect to any finite dimensional linear subspace of C^1.

Therefore, in convex interpolation one should consider nonlinear splines, e.g. exponential splines [7], lacunary splines [3], [8], rational splines [1], [6], [14], or splines on refined grids with variable additional nodes [20]. In the present paper, some of these results will be reviewed in a unified manner, including extensions mainly from a numerical point of view. Here we refer especially to the bounds (14) and (25).

Convex interpolation of $2D$ data has been also of considerably interest. The transfer of the mentioned $1D$ methods seems to be limited so far to the interpolation of gridded data. For applications of tensor product methods we refer to [15], [16]. Only few progress is observed in the convex interpolation of scattered data. There are methods which assure the preservation of convexity only under additional assumptions. But these may be not always satisfied for data in convex position; see [2].

For gridded data in $2D$, the usual convexity is difficult to handle. More adequate seems to be the concept of S-convexity. In the reference [12] this property is introduced by the monotonicity of the first partial derivatives. There can be found also a thoroughly comparison with other convexity definitions. Here we recall that convexity and S-convexity are incomparable, but that S-convexity always implies order-convexity. In $1D$, the mentioned notions are identical. Some of the results concerning the S-convex interpolation of gridded data likewise will be described in what follows.

2. A PARTIALLY SEPARABLE SYSTEM OF INEQUALITIES

In convex interpolation of $1D$ data we are repeatedly led to the following problem. For given constants A_i, B_i, C_i, D_i and τ_i with

$$0 < C_i < A_i,\ 0 < B_i < D_i,\ A_i + B_i = 1,\ C_i + D_i = 1,\ i = 1(1)n; \tag{4}$$

find numbers $p_0, p_1, \ldots, p_n$ which satisfy the partially separable system of inequalities

$$A_i p_{i-1} + B_i p_i \le \tau_i \le C_i p_{i-1} + D_i p_i,\ i = 1(1)n. \tag{5}$$

The solvability can be effectively decided by means of

Algorithm 1 [17]: Define a_i, b_i, $i = 0(1)n$ recursively by

$$\begin{aligned} a_0 &= (\tau_1 - D_1\tau_2)/C_1,\ b_0 = \tau_1,\\ a_i &= \max\{\,\tau_i,\ (\tau_i - C_i b_{i-1})/D_i\,\},\\ b_i &= (\tau_i - A_i a_{i-1})/B_i,\ i = 1(1)n. \end{aligned} \tag{6}$$

Proposition 1 [17]: *Under the assumption (4), problem (5) is solvable if and only if*

$$a_i \le b_i,\ i = 0(1)n, \tag{7}$$

and then all solutions can be determined as follows,

$$\begin{aligned} &p_n \in [a_n, b_n],\ \textit{and for}\ i = n(-1)1:\\ &p_{i-1} \in [\max\{\,a_{i-1}, (\,\tau_i - D_i p_i)/C_i\,\},\ \min\{\,b_{i-1}, (\tau_i - B_i p_i)/A_i\,\}]. \end{aligned} \tag{8}$$

E.g., one can choose the midpoints of these intervals. But in special cases there are better choices; see chapter 4. For a short proof of Proposition 1 we refer to [13]; compare also with [3], [4].

3. CONVEX INTERPOLATION WITH GREGORY'S RATIONAL CUBIC C^1 SPLINES

In the reference [6] Gregory considered the following particular rational cubic splines s on the grid Δ_n. For $i = 1(1)n$ let be

$$s(x) = S_i^T a_i(u) \text{ for } x \in I_i = [x_{i-1}, x_i] \tag{9}$$

with

$$S_i = \begin{pmatrix} h_{i-1} \\ h_i \\ p_{i-1} \\ p_i \end{pmatrix}, a_i(u) = \begin{pmatrix} 1-u \\ u \\ 0 \\ 0 \end{pmatrix} + \frac{u(1-u)}{1+\alpha_i u(1-u)} \begin{pmatrix} 1-2u \\ 2u-1 \\ (1-u)\xi_i \\ -u\xi_i \end{pmatrix}. \tag{10}$$

Here u is a local variable on I_i introduced by $u = (x - x_{i-1})/\xi_i$. The numbers $\alpha_1 \geq 0, \ldots, \alpha_n \geq 0$ are the so-called rationality parameters. We remark that for $\alpha_1 = \ldots = \alpha_n = 0$ cubic splines originate. It is immediately seen that the interpolation condition (3) is incorporated in the representation (9), (10). Further, $s \in C^1$ holds true independently of the choice of the parameters $p_1, \ldots, p_n$, $\alpha_1, \ldots, \alpha_n$, and we get

$$p_i = s'(x_i),\ i = 0(1)n. \tag{11}$$

Finally, s is convex on I if and only if

$$(2+\alpha_i)p_{i-1} + p_i \leq (3+\alpha_i)\tau_i \leq p_{i-1} + (2+\alpha_i)p_i,\ i = 1(1)n. \tag{12}$$

Thus, the solvability of system (12), i.e. of (5) with $A_i = D_i = (2+\alpha_i)/(3+\alpha_i)$, $B_i = C_i = 1/(3+\alpha_i)$ is necessary and sufficient for the existence of convex rational cubic C^1 interpolants. The Algorithm 1 now reads

Algorithm 2 [14]: *Let be*

$$\begin{aligned} &a_0 = (3+\alpha_1)\tau_1 - (2+\alpha_1)\tau_2,\ b_0 = \tau_1 \\ &a_i = \max\{\tau_i, ((3+\alpha_i)\tau_i - b_{i-1})/(2+\alpha_i)\}, \\ &b_i = (3+\alpha_i)\tau_i - (2+\alpha_i)a_{i-1},\ i = 1(1)n. \end{aligned} \tag{13}$$

Next, we are in the position to derive lower bounds for $\alpha_1, \ldots, \alpha_n$ to meet the existence test (7).

Theorem 2: *Let the data set D_n be strictly convex. Then, if the rationality parameters are recursively chosen according to*

$$\begin{aligned} &\alpha_i \geq \max\{0, (\tau_{i+1} + 2a_{i-1} - 3\tau_i)/(\tau_i - a_{i-1})\},\ i = 1(1)n-1, \\ &\alpha_n \geq 0, \end{aligned} \tag{14}$$

the solvability criterion (7) is satisfied.

In order to prove Theorem 2, with (13) we can show inductively that

$$\begin{aligned} &a_0 < \tau_1 = b_0, \\ &a_{i-1} = \tau_{i-1} < \tau_i \leq b_{i-1},\ i = 2(1)n, \\ &a_n = \tau_n < b_n. \end{aligned} \tag{15}$$

Indeed, we can start with $a_0 = \tau_1 - (2+\alpha_1)(\tau_2 - \tau_1) < \tau_1 = b_0$. Further, if $a_{i-1} < \tau_i \le b_{i-1}$ is assumed to be valid for any $i \in \{1, \dots, n-1\}$, then we obtain $((3+\alpha_i)\tau_i - b_{i-1})/(2+\alpha_i) \le \tau_i$ implying $a_i = \tau_i < \tau_{i+1}$, and from (14) it follows that $\tau_{i+1} \le \alpha_i(\tau_i - a_{i-1}) - 2a_{i-1} + 3\tau_i = b_i$. Thus then $a_i < \tau_{i+1} \le b_i$ holds true. Finally, for $i = n$ we get $a_n = \tau_n < \tau_n + (2+\alpha_n)(\tau_n - a_{n-1}) = b_n$.

A weaker form of Theorem 2 is

Corollary 3: *For strictly convex data sets, convex interpolation with Gregory's rational cubic C^1 splines* (9), (10) *is always successful provided that the rationality parameters are sufficiently large.*

We remark that this version extends to the S-convex interpolation of gridded data in $2D$. For the interpolation of scattered data the present rational splines seem to be unsuited. On triangulations already the C^1 property does not occur.

4. CHOICE FUNCTIONALS

If the rationality parameters satisfy (14), the number of convex spline interpolants is infinite. The selection of a suitable spline is usually based on a choice functional. Here we recall two possibilities. The first one is minimum norm modification of an initial approximation $p_0^0, p_1^0, \dots, p_n^0$ for the derivatives in the knots. The corresponding functional is defined by

$$\text{minimize} \sum_{i=0}^{n} (p_i - p_i^0)^2 ; \tag{16}$$

see, e.g., [10]. The second functional

$$\text{minimize} \sum_{i=1}^{n} \frac{w_i}{\xi_i} \{ (p_{i-1} - \tau_i)^2 + (p_{i-1} - \tau_i)(p_i - \tau_i) + (p_i - \tau_i)^2 \} \tag{17}$$

with $w_i = 1$ or $w_i = 1/(1+\tau_i)^{5/2}$ can be considered as an approximation of the mean curvature; see, e.g., [14]. Of course, in both cases the constraints now are the convexity conditions (12).

5. CONVEX INTERPOLATION WITH QUADRATIC C¹ SPLINES ON REFINED GRIDS

In the paper [20] Schumaker considers convex interpolation with quadratic splines on a refined grid $\tilde{\Delta}_n$ of Δ_n characterized by one additional knot

$$\tilde{x}_i = \beta_i x_{i-1} + \alpha_i x_i, \; \alpha_i > 0, \; \beta_i > 0, \; \alpha_i + \beta_i = 1 \tag{18}$$

in each subinterval I_i, $i = 1(1)n$. In convex interpolation the ratios α_i and $\beta_i = 1 - \alpha_i$ have to be variable.

Now, for $i = 1(1)n$ we define the spline s on the subintervals $[x_{i-1}, \tilde{x}_i]$ and $[\tilde{x}_i, x_i]$ by

$$\begin{aligned} s(x) &= S_i^T a_i(u) \quad \text{for} \quad x \in [x_{i-1}, \tilde{x}_i], \\ s(x) &= T_i^T b_i(\mu) \quad \text{for} \quad x \in [\tilde{x}_i, x_i] \end{aligned} \tag{19}$$

with

$$S_i = \begin{pmatrix} h_{i-1} \\ h_i \\ p_{i-1} \end{pmatrix}, \; a_i(u) = \begin{pmatrix} 1-u \\ u \\ 0 \end{pmatrix} + u(1-u)\begin{pmatrix} 1 \\ -1 \\ \alpha_i\xi_i \end{pmatrix}, \tag{20}$$
$$T_i = \begin{pmatrix} h_i \\ \tilde{h}_i \\ p_i \end{pmatrix}, \; b_i(\mu) = \begin{pmatrix} 1-\mu \\ \mu \\ 0 \end{pmatrix} + \mu(1-\mu)\begin{pmatrix} 1 \\ -1 \\ -\beta_i\xi_i \end{pmatrix}.$$

Here u and μ are local variables on $[x_{i-1}, \tilde{x}_i]$ and $[\tilde{x}_i, x_i]$ given by $u = (x - x_{i-1})/(\alpha_i\xi_i)$ and $\mu = (x_i - x)/(\beta_i\xi_i)$, respectively.

Again the interpolation condition (3) is satisfied, and in addition we have

$$\tilde{h}_i = s(\tilde{x}_i), \; i = 1(1)n. \tag{21}$$

Further (11) holds true, and s is from C^1 if and only if the values $\tilde{h}_1, \ldots, \tilde{h}_n$ are chosen such that

$$\tilde{h}_i = \beta_i h_{i-1} + \alpha_i h_i + \frac{\alpha_i\beta_i\xi_i}{2}(p_{i-1} - p_i), \; i = 1(1)n. \tag{22}$$

We remark that this choice is always possible. Therefore, here we have no real C^1 condition. Further, s is easily seen to be convex on I if and only if

$$(1+\alpha_i)p_{i-1} + (1-\alpha_i)p_i \le 2\tau_i \le \alpha_i p_{i-1} + (2-\alpha_i)p_i, \; i = 1(1)n. \tag{23}$$

Thus, in convex interpolation with quadratic C^1 splines on refined grids $\tilde{\Lambda}_n$ we are again led to a problem (5), now with $A_i = (1+\alpha_i)/2$, $B_i = (1-\alpha_i)/2$, $C_i = \alpha_i/2$, and $D_i = (2-\alpha_i)/2$. In this case, Algorithm 1 is specified by

Algorithm 3: *Set*

$$\begin{aligned} a_0 &= (2\tau_1 - (2-\alpha_1)\tau_2)/\alpha_1, \; b_0 = \tau_1 \\ a_i &= \max\{\, \tau_i, \, (2\tau_i - \alpha_i b_{i-1})/(2-\alpha_i) \,\}, \\ b_i &= (2\tau_i - (1+\alpha_i)a_{i-1})/(1-\alpha_i), \; i = 1(1)n \end{aligned} \tag{24}$$

The essential existence test (7) can be always satisfied provided that the ratios $\alpha_1, \ldots, \alpha_n$ are accordingly chosen.

Theorem 4: *We assume that the data set D_n is in strict convex position. Then the solvability criterion (7) holds true if the ratios are taken from the intervals*

$$\begin{aligned} &\alpha_i \in [\max\{\, \epsilon, 1 - 2(\tau_i - a_{i-1})/(\tau_{i+1} - a_{i-1}) \,\}, \, 1), \; i = 1(1)n-1 \\ &\alpha_n \in [\epsilon, 1), \end{aligned} \tag{25}$$

where $\epsilon \in (0,1)$ is fixed. In these cases the convexity of the interpolants (19), (20) is assured.

For proving Theorem 4, also in the present case we can show the inequalities (15). Indeed, at the beginning we obtain $a_0 = \tau_1 - (2-\alpha_1)(\tau_2 - \tau_1)/\alpha_1 < \tau_1 = b_0$. Next, assume $a_{i-1} < \tau_i \le b_{i-1}$ for any $i \in \{1, \ldots, n-1\}$. Then we get $a_i = \tau_i$ because of $(2\tau_1 - \alpha_i b_{i-1})/(2-\alpha_i) \le \tau_i$. In view of (25), for the ratio it holds $1 > \alpha_i \ge 1 - 2(\tau_i - a_{i-1})/(\tau_{i+1} - a_{i-1})$ and thus $\tau_{i+1} \le (2\tau_i - (1+\alpha_i)a_{i-1})/(1-\alpha_i) = b_i$.

Hence we have shown that $a_i = \tau_i < \tau_{i-1} \leq b_i$. For $i = n$ finally, we get $a_n = \tau_n < (2\tau_n - (1+\alpha_n)a_{n-1})/(1-\alpha_n) = b_n$.

In [2] this approach of adding knots is extended to the convex interpolation of scattered data using Powell-Sabin refinements $\tilde{\Delta}$ of an original triangulation Δ. However, in contrast to the univariate case, success cannot always be assured.

6. CONVEX INTERPOLATION WITH RATIONAL LACUNARY C^2 SPLINES

In [19] rational lacunary splines are treated in convex histopolation. These splines allow positive results likewise in convex interpolation. Let now $\alpha_i \geq 0, \beta_i \geq 0$ be rationality parameters and $k_i \geq 3, l_i \geq 3$ lacunarity parameters, $i = 1(1)n$. Further set

$$\varphi(\alpha, k) = 2\alpha^2 + 2\alpha k + k(k-1),\ \psi(\alpha, k) = \alpha + k - 1. \tag{26}$$

For $i = 1(1)n$, we define rational lacunary splines s on the subinterval I_i by

$$s(x) = S_i^T a_i(u) \tag{27}$$

with

$$S_i = \begin{pmatrix} h_{i-1} \\ h_i \\ P_{i-1} \\ P_i \end{pmatrix},\ a_i(u) = \begin{pmatrix} 1-u \\ u \\ 0 \\ 0 \end{pmatrix} + \xi_i^2 \begin{pmatrix} 0 \\ 0 \\ ((1-u)^{k_i}/(1+\alpha_i u) - (1-u))/\varphi(\alpha_i, k_i) \\ (u^{l_i}/(1+\beta_i(1-u)) - u)/\varphi(\beta_i, l_i) \end{pmatrix}. \tag{28}$$

Here the local variable on I_i is given by $u = (x - x_{i-1})/\xi_i$. Now, the interpolation condition (3) is valid, and s is from C^2 if and only if

$$K_i P_{i-1} + L_i P_i + M_i P_{i+1} = \tau_{i+1} - \tau_i,\ i = 1(1)n-1 \tag{29}$$

where

$$K_i = \frac{\xi_i}{\varphi(\alpha_i, k_i)},\ L_i = \frac{\psi(\beta_i, l_i)\xi_i}{\varphi(\beta_i, l_i)} + \frac{\psi(\alpha_{i+1}, k_{i+1})\xi_{i+1}}{\varphi(\alpha_{i+1}, k_{i+1})},\ M_i = \frac{\xi_{i+1}}{\varphi(\beta_{i+1}, l_{i+1})}. \tag{30}$$

Further the C^2 condition (29) assures

$$P_i = s''(x_i),\ i = 0(1)n. \tag{31}$$

Finally, s is convex on I if and only if

$$P_i \geq 0,\ i = 0(1)n. \tag{32}$$

Thus, in convex interpolation with rational lacunary C^2 splines we are interested in the nonnegative solvability of the linear system (29). If we assume for simplicity that

$$k_i = l_i = \alpha_i + 3 = \beta_i + 3 = r + 3, \tag{33}$$

system (29) reads

$$\frac{\xi_i}{r+1}P_{i-1} + 2(\xi_i + \xi_{i+1})P_i + \frac{\xi_{i+1}}{r+1}P_{i+1} = (5r+6)(\tau_{i+1} - \tau_i),\ i = 1(1)n-1. \tag{34}$$

Hence, for sufficiently large r this system has nonnegative solutions provided that D_n is in strict convex position. We remark that the same property holds true if instead of (33), for instance

$$k_i = l_i = \text{ constant } \geq 3,\ \alpha_i = \beta_i = r,\ i = 1(1)n \tag{35}$$

or

$$\alpha_i = \beta_i = \text{ constant } \geq 0,\ k_i = l_i = r,\ i = 1(1)n \tag{36}$$

is assumed. Thus we have obtained

Theorem 5: *Let the data set D_n be in strict convex position. Then the rational lacunary C^2 spline interpolants (27), (28) are convex on I if, e.g. under the assumptions (33), (35), or (36), the parameter r is sufficiently large.*

For determining suitable parameters r the following simple search procedure which terminates after a finite number of steps can be used.

Algorithm 4:

(a) Approximate $P_0 \geq 0$, $P_n \geq 0$ by means of the data set D_n,

(b) choose r, preferably $r = 0$,

(c) solve the tridiagonal diagonally dominant system (34),

(d) check $P_1 \geq 0, \ldots, P_{n-1} \geq 0$; if these inequalities are not valid enlarge r by adding a constant, and goto (c).

Finally we remark that the solvability criterion for convex histopolation from [19] which uses linear programming techniques also applies to the present system (29), (32).

7. S-CONVEX INTERPOLATION WITH RATIONAL BIQUADRATIC C¹ SPLINES

Following the paper [15] we are now concerned with the S-convex interpolation of a $2D$ data set

$$D_{n,m} = \{ (x_i, y_j, h_{i,j}) : i = 0(1)n,\ j = 0(1)m \} \tag{37}$$

on the rectangular grid

$$\Delta_{n,m} :\ x_0 < x_1 < \ldots < x_n,\ y_0 < y_1 < \ldots < y_m. \tag{38}$$

We assume that $D_{n,m}$ is in strict S-convex position, i.e.

$$\begin{gathered} \frac{h_{i,j} - h_{i-1,j}}{\xi_i} < \frac{h_{i+1,j} - h_{i,j}}{\xi_{i+1}},\ i = 1(1)n-1,\ j = 0(1)m, \\ \frac{h_{i,j} - h_{i,j-1}}{\eta_j} < \frac{h_{i,j+1} - h_{i,j}}{\eta_{j+1}},\ j = 1(1)m-1,\ i = 0(1)n, \\ h_{i,j} - h_{i-1,j} - h_{i,j-1} + h_{i-1,j-1} < 0,\ i = 1(1)n,\ j = 1(1)m. \end{gathered} \tag{39}$$

The aim is to construct a C^1 function s on the rectangle $I = [x_0, x_n] \times [y_0, y_m]$ which interpolates $D_{n,m}$,

$$s(x_i, y_j) = h_{i,j}\,,\ i = 0(1)n,\ j = 0(1)m, \tag{40}$$

and which is S-convex on I. That property is defined by the monotonicity of the first partial derivatives on I,

$$\partial_\nu s(x,y) \le \partial_\nu s(\tilde{x},\tilde{y}) \ \text{ for } \ x \le \tilde{x},\, y \le \tilde{y},\ \nu \in \{1,2\}; \tag{41}$$

see [12]. For S-convex interpolation here we use special rational biquadratic splines on the grid $\Delta_{n,m}$. These are defined as follows. For $i = 1(1)n$, $j = 1(1)m$, on the subrectangle $I_{i,j} = [x_{i-1}, x_i] \times [y_{j-1}, y_j]$ set

$$s(x,y) = a_i(u)^T S_{i,j} b_j(v) \text{ for } (x,y) \in I_{i,j}, \tag{42}$$

with the matrix

$$S_{i,j} = \begin{pmatrix} h_{i-1,j-1} & h_{i-1,j} & q_{i-1,j-1} \\ h_{i,j-1} & h_{i,j} & q_{i,j-1} \\ p_{i-1,j-1} & p_{i-1,j} & P_{i-1,j-1} \end{pmatrix} \tag{43}$$

and the vectors $a_i(u) = c(u, \xi_i, \alpha_i)$, $b_j(v) = c(v, \eta_j, \beta_j)$,

$$c(u,\xi,\alpha) = \begin{pmatrix} 1-u \\ u \\ 0 \end{pmatrix} + \frac{u(1-u)}{1+\alpha u} \begin{pmatrix} 1 \\ -1 \\ \xi \end{pmatrix}. \tag{44}$$

Here $u = (x - x_{i-1})/\xi_i, v = (y - y_{j-1})/\eta_j$ are local variables on $I_{i,j}$, and $\xi_i = x_i - x_{i-1}$, $\eta_j = y_j - y_{j-1}$ are mesh sizes. The numbers $\alpha_1 \ge 0, \ldots, \alpha_n \ge 0$, $\beta_1 \ge 0, \ldots, \beta_m \ge 0$ we call rationality parameters.

The interpolation condition (40) obviously holds. Further, the spline s is from C^1 if and only if

$$\begin{aligned} (1+\alpha_i)p_{i,j} + p_{i-1,j} &= \frac{2+\alpha_i}{\xi_i}(h_{i,j} - h_{i-1,j}), \\ (1+\alpha_i)P_{i,j} + P_{i-1,j} &= \frac{2+\alpha_i}{\xi_i}(q_{i,j} - q_{i-1,j}),\ i = 1(1)n,\ j = 0(1)m, \end{aligned} \tag{45}$$

$$\begin{aligned} (1+\beta_j)q_{i,j} + q_{i,j-1} &= \frac{2+\beta_j}{\eta_j}(h_{i,j} - h_{i,j-1}), \\ (1+\beta_j)P_{i,j} + P_{i,j-1} &= \frac{2+\beta_j}{\eta_j}(p_{i,j} - p_{i,j-1}),\ j = 1(1)m,\ i = 0(1)n. \end{aligned} \tag{46}$$

These equalities assure also the continuity of the mixed partial derivative $\partial_1\partial_2 s$ on I, and for $i = 0(1)n$, $j = 0(1)m$ we get

$$p_{i,j} = \partial_1 s(x_i, y_j),\ q_{i,j} = \partial_2 s(x_i, y_j),\ P_{i,j} = \partial_1\partial_2 s(x_i, y_j). \tag{47}$$

System (45), (46) is compatible; for given initial values

$$p_{0,j},\ q_{i,0},\ P_{0,0},\ i = 0(1)n,\ j = 0(1)m \tag{48}$$

it is uniquely solvable.

We remark that, unfortunately, it seems to be impossible to derive C^1 conditions of analogous simplicity if the rationality parameters are local, i.e. if α_i and β_j from the representation (42) - (44) are substituted by $\alpha_{i,j}$ and $\beta_{i,j}$, respectively.

A sufficient condition for the S-convexity of s on I reads

$$\begin{aligned}
&p_{i-1,j} \le p_{i,j}\,,\ i = 1(1)n,\ j = 0(1)m,\\
&q_{i,j-1} \le q_{i,j}\,,\ j = 1(1)m,\ i = 0(1)n,\\
&P_{i,j} \ge 0,\ i = 0(1)n,\ j = 0(1)m,\\
&\xi_i(P_{i-1,j} - P_{i-1,j-1}) + (2+\alpha_i)(q_{i-1,j} - q_{i-1,j-1}) \ge 0,\\
&\eta_j(P_{i,j-1} - P_{i-1,j-1}) + (2+\beta_j)(p_{i,j-1} - p_{i-1,j-1}) \ge 0,\ i = 1(1)n,\ j = 1(1)m.
\end{aligned} \tag{49}$$

Thus, in S-convex interpolation with rational biquadratic splines the solvability of system (45), (46), (49) of linear equalities and inequalities is essential. Again we get solvability for sufficiently large rationality parameters.

Theorem 6 [15]: *Let the data set $D_{n,m}$ be in strict S-convex position. Then, for sufficiently large rationality parameters $\alpha_1, \ldots, \alpha_n$, $\beta_1, \ldots, \beta_m$ the rational biquadratic C^1 spline interpolants* (42) - (44) *are S-convex on the rectangle I.*

For determining suitable parameters, an adapted version of Algorithm 4 can be used.

Finally we remark that the S-convexity is preserved also by particular rational bicubic splines of C^2 continuity; see [18].

REFERENCES

1. J.C. Clements, Convexity-preserving piecewise rational cubic interpolation, SIAM J. Numer. Anal. 27, 1016–1023(1990).

2. J.M. Carnicer and W. Dahmen, Convexity preserving interpolation and Powell-Sabin elements. CAGD 9, 279–289(1992).

3. P. Costantini: On monotone and convex spline interpolation, Math. Comp. 46, 203–214(1986).

4. P. Costantini and R. Morandi, Monotone and convex cubic spline interpolation, Calcolo 21, 281–294(1984).

5. R.L. Dougherty, A. Edelman, and J. M. Hyman, Nonnegativity-, monotonicity-, convexity–preserving cubic and quintic Hermite–interpolation, Math. Comp. 52, 471–494(1989).

6. J.A. Gregory, Shape preserving spline interpolation, Computer-aided design 18, 53–57(1986).

7. W. Heß and J. W. Schmidt, Convexity preserving interpolation with exponential splines, Computing 36, 335–342(1986).

8. P.D. Kaklis and D. G. Pandelis, Convexity-preserving polynomial splines of non-uniform degree, IMA J. Numer. Anal. 10, 223–234(1990).

9. B. Mulansky and M. Neamtu, On the existence of shape preserving interpolation operators, manuscript 1993.

10. B. Mulansky and J. W. Schmidt, Nonnegative C^1 interpolation of scattered data using Powell-Sabin splines, Preprint TU Dresden MATH-NM-02-1993.

11. E. Passow and J. A. Roulier, Monotone and convex spline interpolation, SIAM J. Numer. Anal. 14, 904–909(1977).

12. F.A. Potra, On superadditive rates of convergence, Math. Modell. Numer. Anal. 19, 671–685(1985).

13. J.W. Schmidt, Convex interval interpolation with cubic splines, BIT 26, 377–387(1986).

14. J.W. Schmidt, On shape preserving spline interpolation: existence theorems and determination of optimal splines, Banach Center Publ. 22, 377-389(1989).

15. J.W. Schmidt, Rational biquadratic C^1splines in S-convex interpolation. Computing 47, 87–96(1991).

16. J.W. Schmidt, Positive and S-convex C^1-interpolation of gridded $3D$ data, Preprint TU Dresden MATH NM 03-1993.

17. J.W. Schmidt and W. Heß, Schwach verkoppelte Ungleichungssysteme und konvexe Spline-Interpolation, Elem. Math. 39, 85–95(1984).

18. J.W. Schmidt and W. Heß, S-convex, monotone, and positive interpolation with rational bicubic splines of C^2-continuity, BIT 33(1993).

19. J.W. Schmidt and W. Heß, Shape preserving C^2-spline histopolation, J. Approx. Theory, to appear 1994.

20. L.L. Schumaker, On shape-preserving quadratic spline interpolation, SIAM J. Numer. Anal. 20, 854–864(1983).

2nd Int. Coll. on Numerical Analysis, pp. 187-198
D. Bainov and V. Covachev (Eds)

A FAMILY OF FOUR-STEP EXPONENTIAL FITTED PREDICTOR-CORRECTOR METHODS FOR THE NUMERICAL INTEGRATION OF THE SCHRÖDINGER EQUATION

T.E. Simos and G.V. Mitsou *
Informatics Laboratory, Agricultural University of Athens,
Iera Odos 75, Athens 118 55, Greece.

Abstract

A family of predictor-corrector exponential four-step methods is developed for the numerical integration of the one-dimensional Schrödinger equation. The formula developed contains certain free parameters which allows it to be fitted automatically to exponential functions. The new methods integrate more exponential functions, are very simple compared with the well known sixth algebraic order Runge-Kutta type methods. Numerical results indicate that the new methods is much more accurate than other exponentially fitted methods.

Keywords: Schrödinger equation, predictor-corrector methods, four-step methods, exponential-fitted methods, resonance problem.

1 INTRODUCTION

In recent years the radial Schrödinger equation has been the subject of great activity, the aim being to achieve a fast and reliable algorithm that generates a numerical solution (see [1-4] and [6-13]).

The one dimensional Schrödinger equation has the form:

$$y''(x) = [l(l+1)/x^2 + V(x) - k^2]y(x) \tag{1}$$

where one boundary condition is $y(0) = 0$ with the other boundary condition being specified at $x = \infty$. Equations of this type occur very frequently in theoretical physics, for example [5], and there is a real need to be able to solve them both efficiently and reliably by numerical methods. In

*8 Piliou street, Koliatsou square, 112 55 Athens, Greece

(1) the function $W(x) = l(l+1)/x^2 + V(x)$ is called *the effective potential*, for which $W(x) \to 0$ as $x \to \infty$, and k^2 is a positive real number denoting *the energy*. The boundary conditions are:

$$y(0) = 0 \tag{2}$$

and a second boundary condition, for large values of x, determined by physical considerations.

Boundary value methods based on either collocation or finite differences are not very popular for the solution of (1) due to the fact that the problem is posed on an infinite interval. Initial value methods, such as shooting, need to take into account the fact that $| y'(x) |$ is very large near $x = 0$. So, it is very inappropriate to use standard library packages for the numerical solution of (1).

One of the most popular methods for the solution of (1) is the **Numerov method.** This method is only of order four, but in practice it has been found to have a superior performance to higher order four-step method. The reason for this, as this was proved in [9], is that the Numerov method has the same phase-lag order as the four-step methods but it has a larger interval of periodicity. So, the investigation of linear multistep methods is not a fruitful way to derive efficient high order methods.

An alternative approach to deriving higher order methods for (1) was given by Cash and Raptis [1]. In [1] a sixth order Runge-Kutta type method with a large interval of periodicity was derived. This method has a phase-lag of order six (while the Numerov method has phase-lag of order four) and an interval of periodicity much larger than the method of Numerov.

An alternative approach for developing efficient methods for the solution of (1) is to use exponential fitting. This approach is appropriate because for large x and $k^2 > 0$ the solution of (1) is *periodic*. Raptis and Allison [6] have derived a Numerov type exponentially fitted method. Numerical results presented in [6] indicate that these fitted methods are much more efficient than Numerov's method for the solution of (1). Many authors have investigated the idea of exponential fitting, since Raptis and Allison. Perhaps the most significant work in this general area was that of Ixaru and Rizea [3]. They showed that for the resonance problem defined by (1) it is generally more efficient to derive methods which exactly integrate functions of the form:

$$\{1, x, x^2, \ldots, x^p, \exp(\pm vx), x\exp(\pm vx), \ldots, x^m \exp(\pm vx)\} \tag{3}$$

than to use classical exponential fitting methods. A powerful low order method of this type was developed by Raptis [7]. Also Simos [10] has derived a four-step method of this type which gives much more accurate results compared with other four-step methods. Simos [11] has derive a family of four-step methods which gives more efficient results than other four-step methods. Also Raptis and Cash [8] have derived an exponential fitted method and Cash, Raptis and Simos [2] have derived a method fitted to (3) with $m = 1$ and $p = 3$.

The purpose of this paper is to derive a family of four-step predictor- corrector methods fitted to (3) and in particular to derive methods with $m = 0$ and $p = 7, m = 1$ and $p = 5$ and $m = 2$ and $p = 3$. The new methods are (i) very simple compared with the well known Runge-Kutta type methods. Also they have comparable accuracy, (ii) are much more accurate compared with the exponentially fitted four-step methods proposed by Raptis [12,13] and Simos [10,11]. We have applied the new methods in *the resonance* problem for the one-dimensional radial Schrödinger equation.

2 THE NEW FAMILY OF METHODS

Consider the family of methods:

$$\overline{y}_n = y_n - ah^2(y''_{n+2} - 4y''_{n+1} + 6y''_n - 4y''_{n-1} + y''_{n-2})$$

$$y_{n+2} - 2y_{n+1} + 2y_n - 2y_{n-1} + y_{n-2} = h^2(b_0 y''_{n+2} + b_1 y''_{n+1} + b_2 \overline{y}''_n + b_1 y''_{n-1} + b_0 y''_{n-2}) \tag{4}$$

where, for example, $y''_{n+2} = f(x_{n+2}, y_{n+2})$ with $x_{n+2} = x_n + 2h$.

The local truncation error (LTE) for this method in the classical case ($b_0 = 3/40, b_1 = 13/15, b_2 = 7/60$) is given by

$$LTE = -\frac{h^8(3528ay_n^{(6)} + 95y_n^{(8)})}{30240} + O(h^{10}) \tag{5}$$

We note as " classical case " the case for which b_0, b_1 and b_2 are constant numbers (i.e. the case which is problem independent).

We require that the family of methods (4) should be exact for any linear combination of the functions:

$$Case\ I: \{1, x, x^2, x^3, x^4, x^5, x^6, x^7, \exp(\pm vx)\},$$

$$Case\ II: \{1, x, x^2, x^3, x^4, x^5, \exp(\pm vx), x\exp(\pm vx)\},$$

$$Case\ III: \{1, x, x^2, x^3, \exp(\pm vx), x\exp(\pm vx), x^2\exp(\pm vx)\}. \quad (6)$$

To construct a method of the form (4) which integrates exactly the functions (6), we require that the method (4) integrates exactly:

$$\{1, x, \exp(\pm v_0 x), \exp(\pm v_1 x), \exp(\pm v_2 x), \exp(\pm v_3 x)\} \quad (7)$$

and then put:

$$Case\ I: v_0 = v_1 = v_2 = 0 \text{ and } v_3 = v,$$

$$Case\ II: v_0 = v_1 = 0 \text{ and } v_2 = v_3 = v,$$

$$Case\ III: v_0 = 0 \text{ and } v_1 = v_2 = v_3 = v. \quad (8)$$

The method (4) integrates exactly the functions 1, x. Demanding that (4) integrates exactly (7), we obtain the following system of equations for b_0, b_1, b_2 and a

$$2w_j^2 b_0 \cosh(2w_j) + 2w_j^2 b_1 \cosh(w_j) + b_2 w_j^2 - 2ab_2 w_j^4[\cosh(2w_j) - 4\cosh(w_j) + 3] = 2[\cosh(2w_j) - 2\cosh(w_j) + 1] \quad (9)$$

where $w_j = v_j h, j = 0, 1, 2, 3$

Solving for b_0, b_1, b_2 and a we obtain:

Case I: $v_0 = v_1 = v_2 = 0$ and $v_3 = v$.

$$b_0 = 3/40,\ b_1 = 13/15,\ b_2 = 7/60$$

$$a = \frac{34560[3(40 - 3w^2)\cosh(2w) - 8(13w^2 + 30)\cosh(w) - 7w^2 + 120}{-483840w^4[\cosh(2w) - 4\cosh(w) + 3]} \quad (10)$$

The above formulae are subject to heavy cancellations for small values of $w = vh$. In this case it is much more convenient to use the below mentioned series expansions for the coefficient a of the method.

$$a = \frac{95}{3528} - \frac{83w^2}{105840} + \frac{w^4}{68992} + \frac{577w^6}{10807286400} - \frac{773w^8}{36614882304} \tag{11}$$

Case II: $v_0 = v_1 = 0$ and $v_2 = v_3 = v$.

$$\begin{aligned}
b_0 = \{&384w^3[-26\cosh(w) + 16\cosh(2w) - 6\cosh(3w) + \cosh(4w) + 15] \\
&+192w^4[5\sinh(w) - 4\sinh(2w) + \sinh(3w)] - 32w^5[57\cosh(w) - 30\cosh(2w) \\
&+7\cosh(3w) - 34] + 16w^6[13\sinh(w) + 4\sinh(2w) - 7\sinh(3w)]\}/D \\
b_1 = \{&1536w^3[26\cosh(w) - 16\cosh(2w) + 6\cosh(3w) - \cosh(4w) - 15] \\
&-768w^4[5\sinh(w) - 4\sinh(2w) + \sinh(3w)] - 32w^5[164\cosh(w) - 76\cosh(2w) \\
&+28\cosh(3w) - 7\cosh(4w) - 109] - 64w^6[13\sinh(w) + 4\sinh(2w) \\
&-7\sinh(3w)]\}/D \\
b_2 = \{&-2304w^3[26\cosh(w) - 16\cosh(2w) + 6\cosh(3w) - \cosh(4w) - 15] \\
&+1152w^4[5\sinh(w) - 4\sinh(2w) + \sinh(3w)] - 64w^5[115\cosh(w) - 62 \\
&\cosh(2w) + 13\cosh(3w) + \cosh(4w) - 67] + 32w^6[39\sinh(w) + 12\sinh(2w) \\
&-21\sinh(3w)]\}/D \\
a = \{&192w[15 - 26\cosh(w) + 16\cosh(2w) - 6\cosh(3w) + \cosh(4w)] + 192w^2[5\sinh(w) \\
&-4\sinh(2w) + \sinh(3w)] + 16w^4[13\sinh(w) + 4\sinh(2w) - 7 \\
&\sinh(3w)]\}/(D\ b_2)
\end{aligned} \tag{12}$$

where $D = -192w^5[56\cosh(w) - 28\cosh(2w) + 8\cosh(3w) - \cosh(4w) - 35]$.

The above formulae are subject to heavy cancellations for small values of $w = vh$. In this case it is much more convenient to use the below mentioned series expansions for the coefficients of the method.

$$\begin{aligned}
b_0 &= \frac{3}{40} - \frac{83w^4}{907200} + \frac{w^6}{295680} + \frac{577w^8}{31135104000} \\
b_1 &= \frac{13}{15} + \frac{83w^4}{226800} - \frac{w^6}{73920} - \frac{577w^8}{7783776000} \\
b_2 &= \frac{7}{60} - \frac{83w^4}{151200} + \frac{w^6}{49280} + \frac{577w^8}{5189184000} \\
a &= \frac{95}{3528} - \frac{83w^2}{52920} + \frac{46601w^4}{273829248} - \frac{6328291w^6}{533967033600} + \frac{59172647w^8}{62794523151360}
\end{aligned} \tag{13}$$

Case III: $v_0 = v_1 = v_2 = v_3 = v$.

$$\begin{aligned}
b_0 = \{&64w^3[98\cosh(w) - 64\cosh(2w) + 29\cosh(3w) - 8\cosh(4w) + \\
&\cosh(5w) - 56] - 16w^4[56\sinh(w) - 50\sinh(2w) + 15\sinh(3w) + \sinh(4w) \\
&- \sinh(5w)] + 16w^5[100\cosh(w) - 32\cosh(2w) - 3\cosh(3w) + 5\cosh(4w) - \\
&\cosh(5w) - 69] \quad 16w^6[5\sinh(w) - 4\sinh(2w) + \sinh(3w)] + 16w^7[11\cosh(w) \\
&+4\cosh(2w) - 3\cosh(3w) - 12] + 16w^8[3\sinh(w) - \sinh(3w)]\}/D \\
b_1 = \{&-64w^3[20\cosh(w) - \cosh(2w) - 14\cosh(3w) + 14\cosh(4w) - \\
&6\cosh(5w) + \cosh(6w) - 14] + 16w^4[14\sinh(w) + 40\sinh(2w) - 75\sinh(3w) \\
&+44\sinh(4w) - 9\sinh(5w)] - 16w^5[274\cosh(w) - 80\cosh(2w) - 15\cosh(3w) \\
&+16\cosh(4w) - 3\cosh(5w) - 192] - 32w^6[4\sinh(w) - 6\sinh(2w) + 4\sinh(3w) \\
&- \sinh(4w)] + 64w^7[8\cosh(w) - 6\cosh(2w) + \cosh(4w) - 3] \\
&-64w^8[3\sinh(w) - \sinh(3w)]\}/D \\
b_2 = \{&-128w^3[78\cosh(w) - 63\cosh(2w) + 43\cosh(3w) - 22\cosh(4w) + \\
&7\cosh(5w) - \cosh(6w) - 42] + 32w^4[42\sinh(w) - 90\sinh(2w) + 90\sinh(3w) - \\
&43\sinh(4w) + 8\sinh(5w)] + 32w^5[174\cosh(w) - 48\cosh(2w) - 12\cosh(3w)
\end{aligned} \tag{14}$$

$$+11\cosh(4w) - 2\cosh(5w) - 123] - 16w^6[16\sinh(w) - 16\sinh(2w) + 9$$

$$\sinh(3w) - 4\sinh(4w) + \sinh(5w)] + 16w^7[40\cosh(w) - 8\cosh(2w) + 9$$

$$\cosh(3w) - 4\cosh(4w) - \cosh(5w) - 36] + 96w^8[3\sinh(w) - \sinh(3w)]\}/D$$

$$a = \{24w^2[14\sinh(w) - 20\sinh(2w) + 15\sinh(3w) - 6\sinh(4w) + \sinh(5w)] +$$

$$8w^3[74\cosh(w) - 16\cosh(2w) - 9\cosh(3w) + 6\cosh(4w) - \cosh(5w) - 54] +$$

$$16w^6[3\sinh(w) - \sinh(3w)]\}/(D\, b_2)$$

where $D = -8w^6[42\sinh(w) - 48\sinh(2w) + 27\sinh(3w) - 8\sinh(4w) + \sinh(5w)] + 8w^7[126\cosh(w) - 48\cosh(2w) + 3\cosh(3w) + 4\cosh(4w) - \cosh(5w) - 84]$.

The above formulae are subject to heavy cancellations for small values of $w = vh$. In this case it is much more convenient to use the below mentioned series expansions for the coefficients of the method.

$$\begin{aligned} b_0 &= \frac{3}{40} - \frac{83w^4}{302400} + \frac{2533w^6}{119750400} - \frac{4111w^8}{12454041600} \\ b_1 &= \frac{13}{15} + \frac{83w^4}{75600} + \frac{103w^6}{14968800} - \frac{2921w^8}{778377600} \\ b_2 &= \frac{7}{60} - \frac{83w^4}{50400} - \frac{373w^6}{6652800} + \frac{16949w^8}{2075673600} \\ a &= \frac{95}{3528} - \frac{83w^2}{35280} + \frac{5329w^4}{11409552} - \frac{2198747w^6}{133491758400} + \frac{23467891w^8}{8721461548800} \end{aligned} \qquad (15)$$

3 NUMERICAL ILLUSTRATIONS

In this section we present some numerical results to illustrate the performance of our method. We consider the numerical integration of the Schrödinger equation:

$$y''(x) = (V(x) - k^2)y(x) \qquad (16)$$

in the well-known case where the potential $V(x)$ is the Woods-Saxon potential

$$W(x) = V(x) = u_0/(1+z) - u_0 z/[a(1+z)^2] \qquad (17)$$

with $z = \exp[(x - X_0)/a]$, $u_0 = -50, a = 0.6$ and $X_0 = 7.0$. In order to solve this problem numerically we need to approximate the true (infinite) interval of integration $[0, \infty)$ by a finite interval. For the purpose of our numerical illustration we take the domain of integration as $0 \leq x \leq 15$. We consider (16) in a rather large domain of energies i.e. [1, 1000]. The problem we consider is the so-called *resonance problem*.

In the case of positive energies $E = k^2$ the potential dies away faster than the term $l(l+1)/x^2$; equation (1) effectively reduces to

$$y''(x) + (k^2 - l(l+1)/x^2)y(x) = 0, \tag{18}$$

for x greater than some value X.

The above equation has linearly independent solutions $kxj_l(kx)$ and $kxn_l(kx)$, where $j_l(kx), n_l(kx)$ are the **spherical Bessel and Neumann functions** respectively. Thus the solution of equation (1) has the asymptotic form (when $x \to \infty$)

$$\begin{aligned} y(x) &\simeq Akxj_l(kx) - Bn_l(kx) \\ &\simeq AC[\sin(kx - \pi l/2) + \tan\delta_l \cos(kx - \pi l/2)] \end{aligned} \tag{19}$$

where δ_l is the **phase shift** which may be calculated from the formula

$$\tan\delta_l = \frac{y(x_2)S(x_1) - y(x_1)S(x_2)}{y(x_1)C(x_2) - y(x_2)C(x_1)} \tag{20}$$

for x_1 and x_2 distinct points on the asymptotic region with $S(x) = kxj_l(kx)$ and $C(x) = kxn_l(kx)$.

Since the problem is treated as an initial-value problem, one needs y_0, y_1, y_2 and y_3 before starting a four-step method. From the initial condition, $y_0 = 0$, it is easy to see that, for values of x close to the origin, the solution behaves as $y(x) = cx^{l+1}$ (see for details [9]). To calculate y_2 and y_3 we use the two-step exponentially fitted method of Raptis [7]. With these starting values we evaluate at some point of the asymptotic region the phase shift δ_l and the normalization factor C from the above relations.

For positive energies one has the so-called resonance problem. This problem consists either of finding **the phase shift** $\delta(E)$ or finding those E, for $E \in [1, 1000]$, at which δ equals $\pi/2$. We actually solve the latter problem, known as *" the resonance problem "* when **the positive eigenenergies lie under the potential barrier**.

The boundary conditions for this problem are:

$$\begin{aligned} y(0) &= 0, \\ y(x) &= \cos[\sqrt{E}x] \text{ for large x.} \end{aligned}$$

The domain of numerical integration is [0, 15].

In our numerical illustration we find the positive *eigenenergies or resonances* by the five methods:

Method M I: Method derived by Simos (Algorithm II) [11]

Method M II: Method derived by Simos (Algorithm IV) [11]

Method M III: Method derived by Simos (Case II) [11]

Method M IV: New exponential-fitted method (Case I of the family)

Method M V: New exponential-fitted method (Case II of the family)

Method M VI: New exponential-fitted method (Case III of the family)

The numerical results obtained for the seven methods were compared to the true solution to the Woods-Saxon potential resonance problem. This true solution was obtained correct to six decimal places using the analytical solution. Table 1 shows the absolute errors of the eigenenergies in 10^{-6} units for different choices of constant stepsize, which are shown in column 2. The empty cells indicate that the corresponding absolute errors are larger than 1.

The performance of the different methods is dependent on the choice of the fitting parameter v. For the purpose of obtaining our numerical results it is appropriate to choose v in the way suggested by Ixaru and Rizea [3]. That is, we choose:

The resonance	h	M I	M II	M III	M IV	M V	M VI
53.588872	1/16	405	52	29	12	1	0
	1/32	7	1	0	1	0	0
	1/64	0	0	0	0	0	0
163.215341	1/16	11101	360	197	92	6	0
	1/32	201	6	0	2	0	0
	1/64	3	0	0	0	0	0
341.495874	1/16	192973	1581	134	245	65	6
	1/32	2721	26	9	6	1	1
	1/64	42	1	0	1	0	0
989.701916	1/16			612894	8765	123	9
	1/32	172535	152	80	18	7	2
	1/64	2231	4	2	3	1	0

Table 1: Absolute errors, in 10^{-6} units, of the resonances calculated by the six algorithms M I-M VI

$$v = \begin{cases} (-50 - E)^{1/2} & \text{for } x \in [0, 6.5] \\ (-E)^{1/2} & \text{for } x \in (6.5, 15] \end{cases}$$

For a discussion of the reasons for choosing the values 50 and 6.5 and the extent to which the results obtained depend on these values see [3, pp. 25].

All computations were carried out on an IBM *PC*-AT 80386 with an 80387 mathcoprocessor of the Informatics Laboratory of Agricultural University of Athens using double precision arithmetic (16 significant digits accuracy).

4 Conclusion

We must note that the new methods are very simple (compared with the Runge-Kutta-type hybrid methods of Cash, Raptis and Simos (see [8], [2]) and much more accurate compared with the methods proposed by Raptis [12-13] and Simos [11]. We note that the methods produced by Simos [11] are the most accurate four-step methods in the literature as this has been proved by numerical results shown in [11].

References

[1] J.R. Cash and A.D. Raptis: A high order method for the numerical solution of the one-dimensional Schrödinger. Computer Physics Communications **33**, 299-304(1984).

[2] J.R. Cash, A.D. Raptis and T.E. Simos: A sixth-order exponentially fitted method for the numerical solution of the radial Schrödinger equation. Journal of Computational Physics **91**, 413-423(1990).

[3] L.Gr. Ixaru and M. Rizea: A Numerov-like scheme for the numerical solution of the Schrödinger equation in the deep continuum spectrum of energies. Computer Physics Communications **19**, 23-27(1980).

[4] L.Gr. Ixaru and M. Micu:Topics in Theoretical Physics. Central Institute of Physics, Bucharest, 1978.

[5] L.D. Landau and F.M. Lifshitz: Quantum Mechanics. Pergamon, New York, 1965.

[6] A.D. Raptis and A.C. Allison: Exponential-fitting methods for the numerical solution of the Schrödinger equation. Computer Physics Communications **14**, 1-5(1978).

[7] A.D. Raptis: Two-step methods for the numerical solution of the Schrödinger equation. Computing **28**, 373-378(1982).

[8] A.D. Raptis and J.R. Cash: Exponential and Bessel fitting methods for the numerical solution of the Schrödinger equation. Computer Physics Communications **44**, 95-103(1987).

[9] T.E. Simos: Numerical solution of ordinary differential equations with periodical solution. Doctoral Dissertation, National Technical University of Athens, 1990.

[10] T.E. Simos: A four-step method for the numerical solution of the Schrödinger equation. Journal of Computational and Applied Mathematics **30**, 251-255(1990).

[11] T.E. Simos: Some new four-step exponential-fitting methods for the numerical solution of the radial Schrödinger equation. IMA J. Numer. Anal. **11**, 347-356(1991).

[12] A.D. Raptis: Exponentially-fitted solutions of the eigenvalue Schrödinger equation with auto-

matic error control. Computer Physics Communications **28**, 427-431(1983).

[13] A.D. Raptis: On the numerical solution of the Schrödinger equation. Computer Physics Communications **24**, 1-4(1981).

Bei Fragen zur Produktsicherheit wenden Sie sich bitte an:
If you have any questions regarding product safety,
please contact:

Walter de Gruyter GmbH
Genthiner Straße 13
10785 Berlin
productsafety@degruyterbrill.com